Photonic Technology for Precision Metrology

Photonic Technology for Precision Metrology

Editor

Jacek Wojtas

MDPI • Basel • Beijing • Wuhan • Barcelona • Belgrade • Manchester • Tokyo • Cluj • Tianjin

Editor
Jacek Wojtas
Institute of Optoelectronics
Military University of
Technology
Poland

Editorial Office
MDPI
St. Alban-Anlage 66
4052 Basel, Switzerland

This is a reprint of articles from the Special Issue published online in the open access journal *Applied Sciences* (ISSN 2076-3417) (available at: https://www.mdpi.com/journal/applsci/special_issues/Precision_Metrology).

For citation purposes, cite each article independently as indicated on the article page online and as indicated below:

LastName, A.A.; LastName, B.B.; LastName, C.C. Article Title. *Journal Name* **Year**, *Volume Number*, Page Range.

ISBN 978-3-0365-4493-9 (Hbk)
ISBN 978-3-0365-4494-6 (PDF)

Cover image courtesy of Jacek Wojtas

© 2022 by the authors. Articles in this book are Open Access and distributed under the Creative Commons Attribution (CC BY) license, which allows users to download, copy and build upon published articles, as long as the author and publisher are properly credited, which ensures maximum dissemination and a wider impact of our publications.

The book as a whole is distributed by MDPI under the terms and conditions of the Creative Commons license CC BY-NC-ND.

Contents

About the Editor vii

Jacek Wojtas
Photonic Technology for Precision Metrology
Reprinted from: *Appl. Sci.* **2022**, *12*, 4022, doi:10.3390/app12084022 1

Andrew D. Heeley, Matthew J. Hobbs and Jon R. Willmott
Zero Drift Infrared Radiation Thermometer Using Chopper Stabilised Pre-Amplifier
Reprinted from: *Appl. Sci.* **2020**, *10*, 4843, doi:10.3390/app10144843 9

Hongbo Zhang, Yaping Zhang, Lin Wang, Zhijuan Hu, Wenjing Zhou, Peter W. M. Tsang, Deng Cao and Ting-Chung Poon
Study of Image Classification Accuracy with Fourier Ptychography
Reprinted from: *Appl. Sci.* **2021**, *11*, 4500, doi:10.3390/app11104500 19

Toan-Thang Vu, Thanh-Tung Vu, Van-Doanh Tran, Thanh-Dong Nguyen and Ngoc-Tam Bui
A New Method to Verify the Measurement Speed and Accuracy of Frequency Modulated Interferometers
Reprinted from: *Appl. Sci.* **2021**, *11*, 5787, doi:10.3390/app11135787 33

Janusz Mikołajczyk and Dariusz Szabra
Integrated IR Modulator with a Quantum Cascade Laser
Reprinted from: *Appl. Sci.* **2021**, *11*, 6457, doi:10.3390/app11146457 41

Zbigniew Bielecki, Tadeusz Stacewicz, Janusz Smulko and Jacek Wojtas
Ammonia Gas Sensors: Comparison of Solid-State and Optical Methods
Reprinted from: *Appl. Sci.* **2020**, *10*, 5111, doi:10.3390/app10155111 51

Antoni Rogalski, Piotr Martyniuk, Małgorzata Kopytko and Weida Hu
Trends in Performance Limits of the HOT Infrared Photodetectors
Reprinted from: *Appl. Sci.* **2021**, *11*, 501, doi:10.3390/app11020501 69

Hiroaki Yokota, Atsuhito Fukasawa, Minako Hirano and Toru Ide
Low-Light Photodetectors for Fluorescence Microscopy
Reprinted from: *Appl. Sci.* **2021**, *11*, 2773, doi:10.3390/app11062773 97

About the Editor

Jacek Wojtas

Jacek Wojtas is an associate professor at the Institute of Optoelectronics, Military University of Technology (MUT) in Warsaw, Poland. He obtained a doctorate in technical sciences with honors at the Military University of Technology in 2007. Eight years later, he obtained a postdoctoral degree with honors in the field of technical sciences in the field of electronics based on his scientific achievements and the dissertation "Applications of laser absorption spectroscopy in environmental protection, state security and medicine". He is a member of the IEEE Photonic Society, the Optical Society, and the Committee of Metrology and Scientific Equipment of the Polish Academy of Sciences (a second four-year term since 2016). In 2017, he was elected a member of the Metrology Council appointed by the Minister of Development of the Republic of Poland. He has been managing the Group of the Optical Signals Detection since 2015. He is an academic teacher and expert in optical signal detection, optoelectronic metrology, and optoelectronic sensors for detecting traces of various volatile compounds. His research is focused on the use of state-of-the-art laser technology in optoelectronic gas sensors for environmental applications (gaseous pollution monitoring), medical diagnostics (detecting diseases biomarkers in human breath) and safety systems (explosives detection). He has been awarded many times for his scientific and didactic achievements, including awards from the Minister of National Defense: 1st and 2nd degree, awards of the University Rector, Director of the Institute, and different national awards for the innovation and development of ultra-sensitive sensor technologies. The author or co-author of over 240 scientific publications, an associate or guest editor in three JCR journals and reviewer in over 20 journals and grant agencies.

Editorial
Photonic Technology for Precision Metrology

Jacek Wojtas

Institute of Optoelectronics, Military University of Technology, 00-908 Warsaw, Poland; jacek.wojtas@wat.edu.pl

Abstract: Precision metrology is important for understanding and monitoring various phenomena, especially in special and scientific applications. The subject of the article covers selected aspects of the study of high-resolution measurements of the absorption spectra and their impact on practical issues. This research is crucial for the development of new systems based on laser absorption spectroscopy methods that can detect trace amounts of matter. Some of the most important parameters describing measuring instruments are also defined and illustrated. Specific examples of measurement results are presented, taking into account the need for the highest possible resolution, accuracy and precision.

Keywords: photonic metrology; accuracy; precision; resolution; FTIR; absorption spectroscopy; gas sensors; optoelectronic sensors

1. Introduction

Metrology is a science that covers the issues of measurement related to both experimental and theoretical determinations. It includes the definition and practice realization of measurement units, and reference standards. The most important purpose is to provide a common understanding of measurement principles, not only for research, but for every area of life. The development of technology and applied sciences is very closely related to the development of new and improvement of existing measurement methods. As a result, less measurement uncertainty as well as greater accuracy and precision can be ensured. It manifests itself in increasingly sophisticated scientific discoveries and improvements in everyday devices. What is more, the pursuit of better and better measurements somehow forced the redefinition of International System of Units (SI), which took place in 2019 [1].

Photonics plays an important role in metrology as a technology for the precise measurements of physical and chemical factors ensuring a high level of accuracy, precision and resolution. As the science of photon, it has a decisive influence on the recent scientific and technology achievements. It includes aspects of photon generation, photon-matter interaction and detection. Although it finds many applications in the whole optical range of wavelengths, most solutions operate in the visible and infrared ranges. Since the laser invention, the source of the highly coherent optical radiation, optical measurements have become a perfect tool for highly precise and accurate measurements. Such measurements have additional advantages of being non-contact methods and having fast rates suitable for in-process metrology. However, their extreme precision is ultimately limited by, e.g., the noise of both lasers and photodetectors. Therefore, breakthroughs in the field of optical sources, detectors and optics play a crucial role. They provide the state-of-the-art of photonic technology for precision metrology, and to identify directions for its future development. As a result, the cutting-edge optoelectronic systems are developed and improved in many fields of science and technology (e.g., industry, environment, healthcare, telecommunication, security and space). The most significant areas of photonics applications are:

- Spectroscopy and interferometry;
- Imaging, reflectometry, radiometry photometry and polarimetry;
- In situ, remote and stand-off sensing;
- PICs (photonic integrated circuits), QPICs (photonic quantum sensors) and hybrid sensing;

- Smart sensors and sensors for robots and unmanned systems;
- High-resolution and frequency metrology, and optical clocks.

The importance of photon-based technology, selected principles of metrology and measuring instrumentation in spectroscopy in solving difficult engineering problems in the detection of trace matter are presented in the further and tutorial sections of the article.

2. From the Beginning of Spectroscopy to High-Spectral-Resolution Measurements

Atomic and molecular spectroscopy appeared earlier than quantum mechanics, after Isaac Newton's fundamental study of the spectra of sunlight in 1672, and the demonstration of a narrow slit, instead of a circular hole, for generating spectral lines by William H. Wollaston in 1802 and Joseph Fraunhofer in 1814, who also constructed transmission diffraction grating (an array of slits) and made the first observations of star emission spectra and measurements of dark line wavelengths in the solar spectrum. In 1859, Gustav Kirchhoff (physicist) and Robert Bunsen (chemist) explained that the structure of these spectra is the result of absorption by the components of the Sun's cooler atmosphere of the continuous spectrum emitted by the hotter interior of the Sun. They recognized that each atom and molecule has its own characteristic spectrum (a so-called fingerprint) and established spectroscopy as a scientific tool for analyzing the composition of materials. Another eminent scientist, John F.W. Herschel, proposed the use of continuous curves to illustrate absorption spectra in various media in 1827. Thanks to these and many other achievements, optical spectrometers have found wide application in many fields as a very common analytical tool for the analysis of both terrestrial and stellar objects. Fraunhofer described 574 lines of solar spectrum [2], but, to date, there are over 25,000 lines that have been identified [3].

From the first spectrometer, constructed by Kirchhoff and Benson, to the present day, spectroscopy has become one of the most important tools in the study of matter. Thanks to the development of optomechanical technologies, optical sources and detectors, as well as analog and digital electronics, today's spectrophotometers achieve resolutions that, until recently, were only offered by simulation programs. Figure 1 shows an example of absorption spectra simulated and measured in a mid-infrared range of wavelength, which is very common for gas detection research. There is a comparison of the nitrous oxide (N_2O) absorption spectra obtained in the commonly used HITRAN database and measured with the use of the Bruker IFS 125HR spectrophotometer, which currently offers one of the highest spectral resolutions of absorption spectrum measurements, amounting to 16×10^{-4} cm^{-1}. On the basis of the obtained spectra, it is possible to determine the maximum values of the absorption coefficient for a given wavelength, temperature and pressure. Additionally, it should be noted that, in the oscillatory spectrum, one can distinguish the narrow component lines lying very close to each other, which form the so-called rotational spectrum related to the rotations of molecules. Both types of vibrations, together, constitute the so-called ro-vibronic spectrum. The separation of the rotational spectrum lines was possible, only thanks to the high-resolution measurements. Therefore, the spectra obtained in this way are important from the point of view of the development of optical systems for detecting trace amounts of substances based on laser absorption spectroscopy.

Resolution, apart from accuracy and precision, is one of the most important and most frequently reported metrological factor-characterizing instruments. It is defined as the smallest change that can be measured. As can be observed in Figure 1b, the higher spectral resolution allows the identification of N_2O absorption lines that are invisible at the smaller ones, which is of great importance in the construction of laser gas sensors. This issue will be explained later in the article.

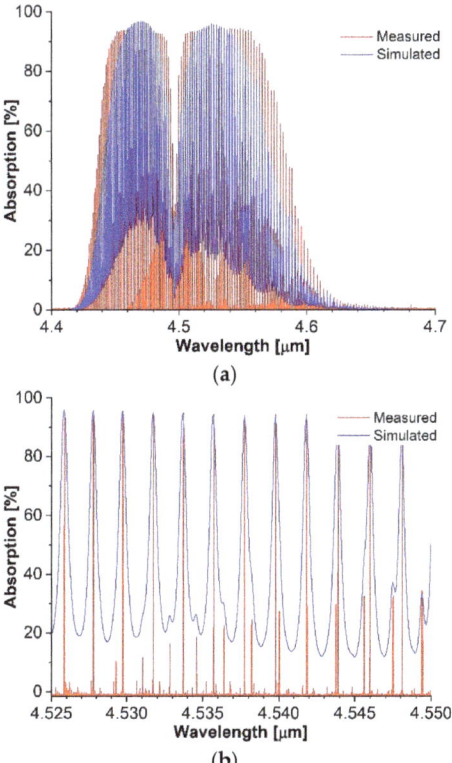

Figure 1. Mid-infrared absorption spectra of nitrous oxide from the HTRAN 2008 database and measured with an IFS 125HR spectrophotometer (**a**), and enlarged part of the spectra to show the differences (**b**). The theoretical spectra were obtained for the following settings: resolution 0.025 cm^{-1} (as a result of the maximum datapoints in the range of 4–5 μm), temperature 296 K, and partial pressure 3.2×10^{-7} atm. The measurement was carried out in the following conditions: resolution 0.0035 cm^{-1}; 10 cm gas cell, including CaF$_2$ windows; 10 mbar; temp. 23 °C; beam splitter CaF$_2$; and liquid-nitrogen cooled InSb detector.

Accuracy, in turn, can be described as a certain amount of measurement uncertainty, with respect to an absolute standard. In other words, this means that an accurate instrument will provide measurements closest to the true value or standard (low uncertainty). Precision describes the reproducibility of the measurement. Accuracy is sometimes confused with precision; that is, the ability of a device to make consecutive, close measurements. The instrument can produce extremely precise measurements that are not at all accurate. Precision is a direct result of the repeatability and reproducibility of the measurements. The differences between these parameters are shown in Figure 2.

The manner by which the accuracy and precision of the measuring instrument, in this case a spectrophotometer, influences the results is presented.

In the presented definitions, there is a term that is very important for all measurements, uncertainty, which describes the doubt concerning the validity of the measurement result. It generally characterizes a dispersion of values that can reasonably be assigned to the measurand, such as, for example, standard deviation or the half-width of an interval with a certain confidence level [4]. In the presented example, the standard deviation (σ_x), calculated from Formula (1), is 2.51×10^{-5} µm.

$$\sigma_x = \sqrt{\frac{1}{N-1} \sum (x_i - \bar{x})^2} \qquad (1)$$

where N is the number of measurements, x_i—results of the subsequent measurements, and $\bar{x}$ is the mean value.

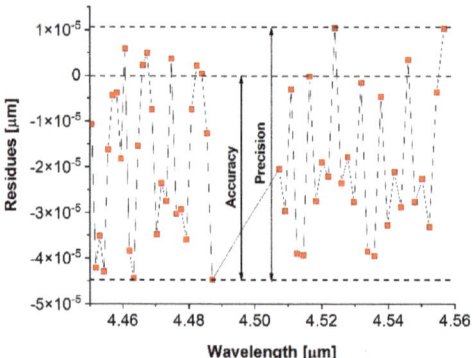

Figure 2. Plot of differences between the centers of the 53 highest N_2O absorption lines in the range of 4.4–4.7 µm, determined on the basis of calculations from the HITRAN database and measured with IFS 125HR.

3. Some Practical Aspects of High-Resolution Measurements of Absorption Spectra

Nitrous oxide is primarily known as a laughing gas and a gas used to improve the power of car engines (during racing). However, the most important motivation is that N_2O is an explosives marker. It is used to assess their condition (aging process) as well as for detection [5,6]. From a chemical point of view, most of the modern explosives are organic nitro compounds of aromatic, heterocyclic or aliphatic series, having one or more nitro groups linked to the molecular skeleton through carbon (C-NO_2), nitrogen (N-NO_2) or oxygen (O-NO_2). It is well known that, as a result of the thermal decomposition of nitrates and nitro compounds, nitrogen oxides are formed (including N_2O), and therefore an indirect method of explosives detection may be the analysis of gases released during their heating. The composition of gaseous products of thermolysis, including the amount and type of nitrogen oxides, depends primarily on the conditions of its operation (e.g., temperature and state of aggregation), but also on the structure of the explosive compound undergoing decomposition, i.e., on the number of nitro groups and the method of their connection with the skeleton carbon molecule. For this reason, the results of the analysis can also be used to identify the compound to be degraded (or at least a group of compounds) [7,8]. Extensive research on the decomposition products of various explosives, where N_2O has been identified as a common marker, is described in refs. [6,9]. Figure 3 shows the examples of the research results conducted in this area at the Institute of Optoelectronics MUT. The graph shows the N_2O spectra recorded during the decomposition of propylnitroguanidine (PrNQ), which is a future TNT substitute.

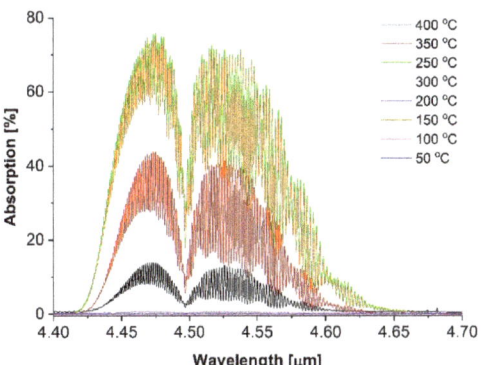

Figure 3. Exemplary N$_2$O absorption spectra observed in the vapors obtained from a 2 mg PrNQ sample at a temperature of 50 to 400 °C. The measurement was carried out at the following conditions: resolution 0.5 cm^{-1}, 2.4 m gas cell, 660 mbar, beam splitter KBr, and liquid nitrogen cooled MCT detector.

Taking this into account, it makes sense to develop sensors capable of detecting N$_2$O. It should also be taken into consideration that these are trace concentrations. and therefore the sensor must have the highest possible sensitivity and excellent parameters, which were previously described. Therefore, the sensors that use laser absorption spectroscopy (LAS) are ideal for this purpose [8,9]. The fast response, non-invasive measurement and high reliability of these sensors can be highlighted alongside their other advantages. However, in order to achieve such features, one of the fundamental conditions that must be met is to match the wavelength of the radiation source to the absorption spectrum of the test substance. Therefore, lasers that have narrow spectral lines (e.g., FWHM < 0.001 cm^{-1}) are used. As a result, a precise adjustment of the laser radiation wavelength to the selected absorption bands (lines), the half-width of which is also often below 0.001 cm^{-1}, can be achieved. Moreover, in order to ensure the possibility of detecting the lowest possible concentration of the molecules of a given substance, the bands with the largest absorption coefficient are selected. However, to ensure the high accuracy of the detection system, the interference of the absorption bands of other substances in the vicinity of the analyzed sample should also be taken into account. As can be observed in Figure 1b, in the case of low-resolution absorption spectra, it may occur that the laser is tuned between the absorption lines due to the apparent shape of the absorption lines formed as envelopes from the points included in different and independent absorption lines.

The accurate and precise measurements of the absorption spectra have another very important meaning. If the sensor is not perfectly tuned at the maximum of the absorption line, not only will it decrease its sensitivity, but it will also be exposed to a highly non-linear operation. Some well-known equations are needed to explain this. Absorption gas sensors use the effect of the absorption of radiation by molecules, the wavelength (λ) of which is matched to their resonance frequency, the so-called absorption band. The radiation $I(\lambda)$ passing through the layer of gas (L) with the absorption coefficient $\alpha(\lambda)$ is attenuated according to the Beer–Lambert law:

$$(\lambda) = I_0(\lambda)e^{-\alpha(\lambda)L}, \qquad (2)$$

where $I_0(\lambda)$ is the initial intensity, before the absorbing layer. Absorbance (*Abs*) can be used to define the quantity characterizing the absorption of radiation, which is described by the following formula:

$$Abs = \alpha(\lambda)L = \sigma(\lambda)NL = \ln\left(\frac{I_0(\lambda)}{I(\lambda)}\right), \qquad (3)$$

where $\sigma(\lambda)$ denotes the absorption cross section in cm^2, and N is the absorber concentration in cm^{-3}. In the case of the inaccurate matching of the laser wavelength to the absorption line or using not single-mode laser, then the absorption coefficient does not possess the same value. This is illustrated in Figure 4. For the sake of simplicity, the same spectral full widths in the half maximum ($FWHM_1 = FWHM_2$) of the radiation sources were assumed.

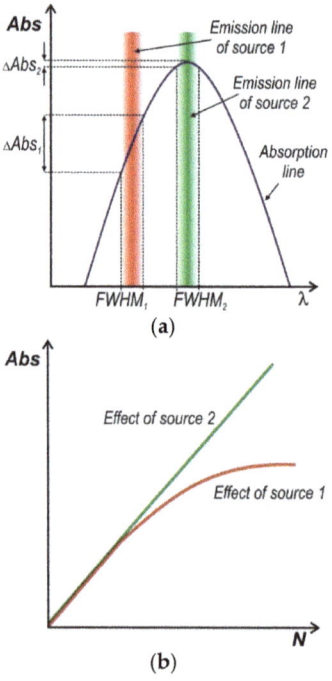

Figure 4. Illustration of the imprecise matching of the emission line of the radiation source to the maximum of the absorption line (**a**) and the effect on the absorbance (**b**).

Figure 4a shows that, for the first source, small changes in the wavelength cause significant changes in the absorbance (ΔAbs_1). On the other hand, in the case of source 2, these changes are much smaller ($\Delta Abs_2 \ll \Delta Abs_1$). It follows that such an arrangement is much less sensitive to the shape of the absorption line. As the absorber concentration increases, the shape of the absorption line changes (its amplitude and width increase). This results in the non-linear response of the absorption sensor with source 1 (Figure 4b).

4. Summary

Photonics plays a key role in increasing the precision of measurements in many areas of life and science. The present article justifies this thesis on the example of evolution and some principles of photon spectroscopy. Since the inception of spectroscopy, to which such eminent scientists as Newton, Wollaston, Fraunhofer, Kirchhoff, and Herschel have contributed, this field of research has come a long way to produce advanced devices that can be used today. They offer a high spectral resolution and precise, accurate, non-invasive and fast measurements. The meaning of these parameters is presented in the present article. As this paper demonstrates, they have a direct impact on practical applications. Measurements, with the use of such spectrophotometers, make it possible to record the rotating structure of the oscillating spectrum of the characteristic absorption bands of various substances. Thanks to this, in gas absorption sensors, it is possible to precisely match the wavelength of the laser radiation to the selected spectral line, where the absorption coefficient is the highest. With the development of technology, both in the field of semiconductor radiation sources and highly responsive detectors, there are opportunities to develop improved and miniaturized sensors based on absorption analysis and in previously unused spectral ranges.

Funding: The presented results were obtained within the project funded by the Ministry of National Defense (PL), grant no. GBMON/13-993/2018/WAT.

Institutional Review Board Statement: Not applicable.

Informed Consent Statement: Not applicable.

Data Availability Statement: Not applicable.

Conflicts of Interest: The author declares no conflict of interest.

References

1. The International System of Units (SI). Available online: https://www.bipm.org/en/measurement-units (accessed on 31 March 2022).
2. Hentschel, K. *Mapping the Spectrum: Techniques of Visual Representation in Research and Teaching*; Oxford University Press: Oxford, UK, 2002.
3. Fraunhofer Lines. Available online: https://www.britannica.com/science/Fraunhofer-lines (accessed on 31 March 2022).
4. Evaluation of Measurement Data—Guide to the Expression of Uncertainty in Measurement, JCGM 100:2008 GUM 1995 with Minor Corrections. Available online: https://www.bipm.org/documents/20126/2071204/JCGM_100_2008_E.pdf/cb0ef43f-baa5-11cf-3f85-4dcd86f77bd6 (accessed on 31 March 2022).
5. Wojtas, J.; Bogdanowicz, R.; Kamienska Duda, A.; Pietrzyk, B.; Sobaszek, M.; Prasuła, P.; Achtenberg, K. Fast-response optoelectronic detection of explosives' residues from the nitroaromatic compounds detonation: Field studies approach. *Measurement* **2020**, *162*, 107925. [CrossRef]
6. Wojtas, J.; Stacewicz, T.; Bielecki, Z.; Rutecka, B.; Medrzycki, R.; Mikolajczyk, J. Towards optoelectronic detection of explosives. *Opto-Elektron. Rev.* **2013**, *21*, 210–219. [CrossRef]
7. Klapotke, T.M. *Chemistry of High-Energy Materials*; Walter de Gruyter: Berlin, Germany, 2011.
8. Oxley, J.C. A survey of the thermal stability of energetic materials. In *Energetic Materials, Decomposition Crystal and Molecular Properties*; Politzer, P., Murray, J.S., Maksic, Z.B., Eds.; Elsevier: Amsterdam, The Netherlands, 2003.
9. Wojtas, J.; Szala, M. Thermally enhanced FTIR spectroscopy applied to study of explosives stability. *Measurement* **2021**, *184*, 11000. [CrossRef]

Article

Zero Drift Infrared Radiation Thermometer Using Chopper Stabilised Pre-Amplifier

Andrew D. Heeley, Matthew J. Hobbs and Jon R. Willmott *

Portobello Centre, Sensor Systems Group, Electronic & Electrical Engineering Department, University of Sheffield, Pitt Street, Sheffield S1 4ET, UK; adheeley1@sheffield.ac.uk (A.D.H.); m.hobbs@sheffield.ac.uk (M.J.H.)
* Correspondence: j.r.willmott@sheffield.ac.uk; Tel.: +44-114-222-5436

Received: 8 June 2020; Accepted: 11 July 2020; Published: 15 July 2020

Abstract: A zero-drift, mid–wave infrared (MWIR) thermometer constructed using a chopper stabilised operational amplifier (op-amp) was compared against an identical thermometer that utilised a precision op-amp. The chopper stabilised op-amp resulted in a zero-drift infrared radiation thermometer (IRT) with approximately 75% lower offset voltage, 50% lower voltage noise and less susceptibility to perturbation by external sources. This was in comparison to the precision op-amp IRT when blanked by a cover at ambient temperature. Significantly, the zero-drift IRT demonstrated improved linearity for the measurement of target temperatures between 20 °C and 70 °C compared to the precision IRT. This eases the IRT calibration procedure, leading to improvement in the tolerance of the temperature measurement of such low target temperatures. The zero-drift IRT was demonstrated to measure a target temperature of 40 °C with a reduction in the root mean square (RMS) noise from 5 K to 1 K compared to the precision IRT.

Keywords: infrared thermometer; mid-wave infrared; indium arsenide antimony photodiode; uncooled thermometer; fibreoptic coupling; chopper stabilised op-amp; zero-drift pre-amplifier

1. Introduction

Non-contact temperature measurements, acquired from processes using infrared radiation thermometers (IRTs), afford advantages over contact temperature measurements. These include amelioration of the susceptibility of thermocouple wires to degradation from exposure to the measurement conditions and the effects of contact upon the measurand and object [1,2]. Thermocouples afford a wide range of temperature measurements, typically having the capability of continuous measurement from zero to many hundreds of degrees Celsius [1]; a capability that would need to be replicated by IRT based thermocouple replacements. IRTs that use mid-wave infrared (MWIR) detectors can be deployed in various applications that need the capability to measure lower temperatures, for example, measurement of temperatures below 100 °C. Mercury Cadmium Telluride (MCT) photon detectors have been used in low-temperature IRTs but require thermoelectric cooling to reduce their dark current and Johnson noise, whilst increasing their photosensitivity to enable the measurement to be acquired [3]. Cooling a photon detector increases the cost and size of IRTs, making their usage undesirable, particularly within size-constrained instrumentation [4]. Thermal detectors, such as thermopiles, can measure over this spectral range, however, they suffer from slow response time and low sensitivity compared to photon detectors [1]. Thermal detectors require high gain amplification to achieve output voltages that can be measured with benchtop instrumentation and can only represent, accurately, signals that change slowly, as demonstrated by Moisello et al. [5]. Both MCT and thermal detector technologies incur drift that increases the measurement uncertainty and introduces non-linearity. Variation of voltage offset with IRT ambient temperature is of particular concern when using uncooled detectors [6]. A mechanical chopper, or shutter, can be used to overcome drift within IRTs to improve their stability [7,8] but at the expense of increasing the overall instrument cost and size. In addition, IRTs incorporating mechanical chopping also need detectors with lower shunt capacitance. This enables them to keep up with the chopper which, presumably, is higher frequency than any measurement expected of the IRT.

Typically, IRTs that use photodiodes also utilise a transimpedance amplifier (TIA) to provide a measurable voltage from the photocurrent generated. MWIR photodiodes generate small magnitude photocurrents when exposed to low radiant flux, which requires a high transimpedance to generate a measurable signal from the TIA. This dictates the need for op-amps with low noise and input voltage offset to be used in the TIA to afford the measurement of low temperatures with low uncertainty. Manufacturers offer 'precision' op-amps that have low magnitudes of input voltage offset, input current bias, noise and temperature-induced zero offset drift. Precision op-amps are often used in amplifier circuits to achieve precision measurements from low magnitude signals [9–11]. Similarly, low magnitude voltages to those arising from MWIR photodiodes but arising from physiological, micro-electro-mechanical-systems (MEMS) sensors and thermal detectors, have been demonstrated to be measurable using high gain amplifiers that had low noise and offset [5,12–16].

An additional consideration is how the variation of the ambient temperature can lead to drift in the output voltage of typical TIA circuits used in IRTs, causing non–linearity in the IRT response and increased measurement uncertainty, which needs to be mitigated to allow low magnitude currents and voltages to be measured accurately [4,17,18]. These limitations in IRT characteristics have constrained the capability of MWIR thermometers to achieve low-temperature measurements with low uncertainties when using uncooled detectors, even with the use of a precision op-amp.

A particular type of op-amp that offers improved performance over a precision op-amp is the chopper stabilised or zero-drift, op-amp. Typically, chopper stabilised op-amps incur very low input offset voltage, achieving an order of magnitude lower offset voltage than precision op-amps. Equally, the temperature-related drift in this offset voltage is typically three orders of magnitude lower for chopper-stabilised op-amps compared to precision op-amps. Both precision and chopper-stabilised op-amps afford sufficiently low noise magnitude to enable small signals to be amplified to measurable magnitudes with signal-to-noise ratios greater than unity [9,10,19,20]. Chopper-stabilised op-amps modulate the input voltage with a square wave carrier signal and demodulate the resulting signal to isolate the output voltage from low-frequency noise [20]. This modulation and demodulation occur within the op-amp's integrated circuit, using transistor switches to afford kilohertz modulation frequencies. This modulation method results, ultimately, in the very low offset voltage and offset voltage drift of the chopper stabilised op-amp output voltage [20–22].

Chopper-stabilised op-amps have been used for the measurement of signals within various applications, including signals from physiological potential sensors, micro-electro-mechanical systems (MEMS) gas measurement sensors and thermopiles in presence detectors. Moisello et al., Butti et al., Kim et al., Wang et al., and Han et al., have demonstrated the application of chopper stabilised amplifiers to the acquisition of measurements from sensors having low output signal magnitudes [5,13–15,23]. Butti et al., and Lee et al., have further demonstrated the suitability of applying chopper stabilised op-amps to the amplification of signals from low signal sensors with elevated impedance to achieve measurements with low noise and offset [23,24]. Chopper-stabilised op-amps have been demonstrated to be capable of amplifying weak sensor output voltages, of the order of microvolts, with sufficient signal to noise ratio to enable extraction of measurements [5,12,13,15,25,26]. Given the benefit provided by chopper stabilised op-amps for such applications, it would be reasonable to expect them to also offer benefits within an IRT incorporating an uncooled MWIR photodiode. Use of a chopper stabilised op-amp within such an IRT could enable the measurement of lower minimum temperatures, associated with low magnitude photocurrents generated and voltages amplified from them, without compromising the IRT response time, sensitivity, measurement uncertainty or range of the temperature measurement.

A fibreoptic coupled MWIR thermometer is presented that comprised a bootstrapped Indium Arsenide Antimony (InAsSb) photodiode connected to a novel TIA circuit that utilised a chopper stabilised op-amp. The usage of a chopper-stabilised op-amp within the TIA should afford lower input voltage offset, noise and offset voltage drift than an equivalent IRT using a precision op-amp, thereby affording lower minimum temperatures to be measured. The noise and offset voltage of the resulting 'zero-drift' IRT was compared to a similar MWIR thermometer comprising an identical circuit, except that the TIA used a precision op-amp. The comparison between the IRTs was used to infer whether the advantages expected from the chopper stabilised op-amp IRT was demonstrable in laboratory-based measurements. Performing analyses without any lens optics, both with and without the fibreoptic in place, enabled a direct comparison of the IRTs to be undertaken, without the uncertainties associated with optical alignment. In this work, we measure the temperature of calibrated blackbody sources in units of degree Celsius, °C. However, for subsequent analysis of instrument noise performance and resolution, we differentiate performance from a position within the temperature scale by using the different unit of kelvin, K, where 1 K = 1 °C.

2. Materials and Methods

Two IRTs were configured in order to directly compare and contrast a zero-drift IRT against a precision IRT. The InAsSb photodiode used within each IRT was an uncooled Hamamatsu P13243-011MA (Hamamatsu Photonics, Hertfordshire, UK), with a quoted specific detectivity (D^*) and peak responsivity values of 1×10^9 cm·Hz$^{\frac{1}{2}}$/W and 4.5 mA/W, respectively. No additional filter was placed in front of the detector, ensuring a broadband response defined by the detector spectral sensitivity. Each IRT comprised an identical TIA circuit, except for the type of op-amp used. A Linear Technologies LTC2050 chopper stabilised op-amp (Linear Technologies, Milpitas, CA, USA) and a Linear Technologies LT1012 precision op-amp (Linear Technologies, Milpitas, CA, USA) were used for the zero-drift and precision IRTs, respectively. The TIA feedback network resistance and capacitance were 10 MΩ and 10 pF, leading to an RC time constant of 100 µs. This provided a faster response time than could be achieved by thermocouples or thermal detector based IRTs, whilst still providing sufficient low-pass filtering. The circuits were not shielded, therefore, enabling some degree of electromagnetic coupling (EMC) related pickup to affect both IRTs to similar extents.

2.1. Comparison of IRTs Measuring Steady-State Temperature

Comparisons were undertaken to determine the capabilities of the IRTs to maintain a nominal zero value when the IRTs were blanked by a cover at ambient temperature and a steady-state value when the photodiodes were exposed to a source at an arbitrary steady-state temperature. The IRTs were exposed to radiant sources with no optical configuration, other than the intrinsic window of the photodiodes. This, therefore, ensured a fair and direct comparison between the IRTs without the usual uncertainties associated with optical alignment, including aligning both IRTs to the target source at the same time. The IRTs were positioned at the same arbitrary distance of 150 mm from the radiant source and adjacent to the aperture centreline. The IRTs, therefore, measured the same radiant source concurrently and from adjacent positions, thereby ensuring that no difference arose from systematic errors associated with the experimental configuration. The TIA output voltages were recorded on a pair of Keysight Technologies 34465A digital multi-meters (DMMs) (Keysight Technologies Malaysia, Penang, Malaysia) with identical configurations to measure and store data. Measurements were recorded at 1 min intervals for three consecutive durations of 100 h each. The data sets were concatenated to provide contiguous IRT output voltage measurements over a total duration of 300 h. The configuration of the blackbody source and photodiodes is illustrated in Figure 1.

The output voltages measured from the IRTs under the zero signal conditions were interpreted as resulting from the total of the offset voltage and noise of the op-amps only. The source was changed from providing a zero signal condition to both of the IRTs receiving blackbody radiation from an Ametek Land Instruments Landcal P1200B blackbody source (Ametek Land Instruments, Dronfield, UK), set at an arbitrary and constant temperature of 700 °C. The noise power on the output voltage was represented by the variance around the mean in both zero signal and steady-state signal tests.

It was not critical that identical conditions were achieved for each IRT because small differences between the mean output voltages would not influence the variances. The excess kurtosis values of the measurements were evaluated to indicate whether the distributions had extensive or restricted tails and to indicate how much extremal values affected the variances.

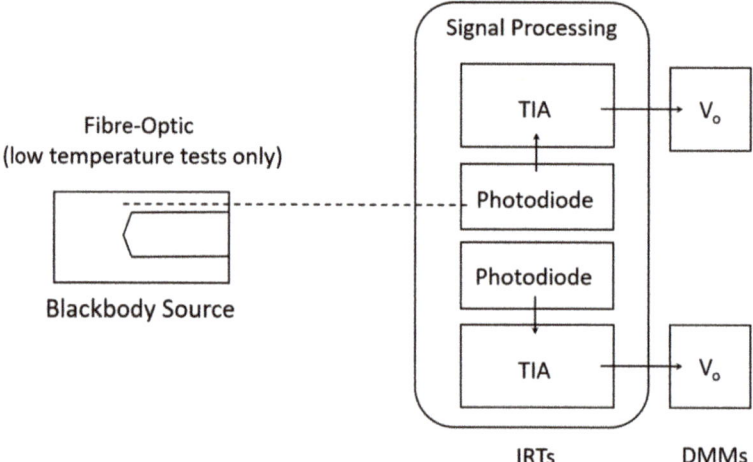

Figure 1. Schematic diagram illustrating configuration used to test the two infrared radiation thermometers (IRTs) concurrently.

2.2. Characterisation of IRTs over Low-Temperature Ranges

The output voltages of the IRTs were characterised as a function of the temperature of an Ametek Land Instruments, Landcal P80P blackbody source (Ametek Land Instruments, Dronfield, UK). The blackbody source temperature was calibrated prior to characterisation of the IRTs, using an Isothermal Technology Limited, milliK precision thermometer (Isothermal Technology, Southport, UK) and transfer standard Platinum Resistance Thermometer (Isothermal Technology, Southport, UK), which was calibrated at manufacture by an accredited, primary temperature calibration laboratory.

A 150 mm length of 600 μm core diameter, ZBLAN fluoride glass fibreoptic coupled blackbody radiation from the calibration thermowell in the blackbody source to the IRTs. The fibreoptic had a numerical aperture of 0.20, transmitting radiation between 0.2 μm and 4.5 μm in wavelength. The depth to diameter ratio of the calibration thermowell and the exposed tip of the fibre exceeded 5:1, which afforded high relative emissivity approaching blackbody conditions [27,28].

The IRT output voltages were measured and recorded using a DMM. The temperature set point of the blackbody source was increased in 10 °C increments between 20 °C and 70 °C and the IRT output voltages characterised against this sequence of temperatures. The IRT output voltages at each source temperature were recorded at intervals of 1 s for a duration of ten minutes per measurement. The configuration of the equipment for undertaking IRT characterisation was the same as presented in Figure 1, except that the fibreoptic was inserted into the blackbody source thermowell and coupled the blackbody radiation onto the photodiode.

The mean and standard deviation of the output voltages recorded from the IRTs were evaluated to abstract values that were representative of each temperature and the noise. The root mean square (RMS) noise for each IRT was calculated from the characterisation measurements.

The Wien approximation of Planck's law enabled evaluation of the ratio between radiances arising from two temperatures to be compared. This leads to the expectation of a linear relationship between the natural logarithm of the IRT output voltage, $ln(V_o)$ and the inverse of the absolute temperature of the source object, $1/T$. The limiting effective wavelength, which is a monochromatic wavelength abstraction of the IRT response at different temperatures over a broad wavelength range [29], was calculated from the gradient of the relationship between $ln(V_o)$ and $1/T$. The practical calibrations of the IRTs were used to determine interpolation curves, according to the method of Sakuma and Hattori [30]. RMS noise was evaluated from the calibration data to provide an IRT noise measurement in terms of the equivalent temperature resolution in K.

These measurements tested the capability of the IRTs to measure low temperatures likely to be encountered on industrial processes [31] and provided practical calibrations of the IRTs through the temperature range 20 °C to 70 °C.

3. Results and Discussion

3.1. Comparison of the Capabilities of the IRTs to Maintain 'Zero' and Steady-State Output Voltages

The capability of the IRTs to maintain steady-state output was determined based upon the variability of output voltage measurements recorded under two separate measurement conditions, as shown in Figure 2. These measurements were performed with (a) the IRTs blanked with a cover at ambient temperature and (b) the IRTs sighted on the blackbody source radiating at an arbitrary steady-state temperature of 700 °C.

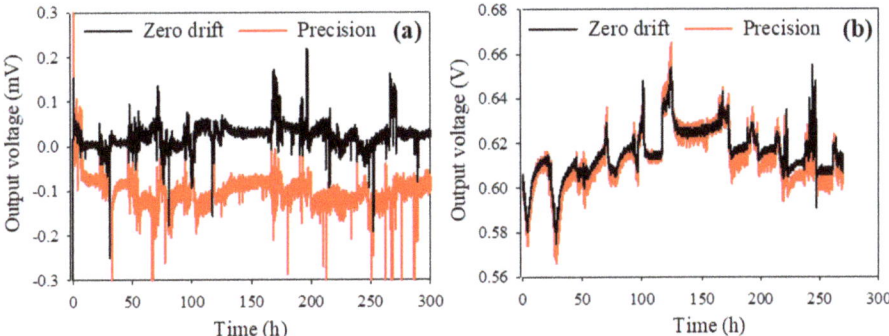

Figure 2. IRT output voltage comparison with (**a**) IRTs blanked by the cover at ambient temperature and (**b**) IRTs sighted upon blackbody source radiating at 700 °C.

The variances for the zero-drift IRT can be seen to be smaller than the variances incurred by the precision IRT; these variances were calculated to be 8.6×10^{-4} (mV)^2 and 19.1×10^{-4} (mV)^2, respectively, therefore representing over 50% reduction in voltage noise. It was also evident by inspection of Figure 2a that the zero-drift IRT maintained the nominal 'zero' output voltage closer to zero millivolts than the precision IRT. The mean output voltages were +0.024 mV and −0.10 mV, respectively, therefore representing a 75% reduction in offset voltage. The precision IRT incurred a larger magnitude drift from zero- than the zero-drift IRT; the drift and cycling incurred by the precision op-amp IRT mostly occurred within the range −0.05 mV to −0.10 mV during the blanking tests. In comparison, the zero-drift IRT incurred variations mostly between 0.00 mV and +0.05 mV. Both IRTs demonstrated some larger excursions that occurred randomly during the tests, although the precision IRT incurred more excursions than the zero-drift IRT. It is believed that these excursions were caused by external interference sources, suggesting that the zero-drift IRT was more robust against EMC related pickup. Both IRTs demonstrated larger magnitude changes in the output voltage than the noise incurred around the mean values. These had similar magnitudes and occurred concurrently during the steady-state 700 °C set-point temperature tests, as illustrated by inspection of Figure 2b. The similarity of magnitude and occurrence of these output voltage changes for both IRTs implied that the cause was common. The upwards and downwards trend observed in the output voltage of both IRTs is believed to be due to the red-shift in the cut-off wavelength of the photodiodes. This shift occurs due to drift in the ambient temperature, which is identical for each IRT.

3.2. Statistics Describing the Distributions of Measurements

Sample tests demonstrated that the measurements acquired from the two IRTs at the zero signal and 700 °C conditions arose from independent systems. Therefore, any differences between the mean values, variances and excess kurtosis values calculated were interpreted as arising from the different op-amps used to configure the TIAs.

The mean, variance and excess kurtosis values calculated for the IRT output voltage measurements under the two measurement conditions are presented in Table 1.

Table 1. Mean, variance and excess kurtosis values of precision and zero-drift IRTs.

Condition	IRT Version	Mean V_o (V)	Variance (V^2)	Excess Kurtosis
Zero Signal	Precision	-1.000×10^{-4}	1.906×10^{-9}	416.2
	Zero-Drift	2.397×10^{-5}	8.558×10^{-10}	121.4
700 °C Signal	Precision	0.613	1.645×10^{-4}	1.22
	Zero-Drift	0.616	1.168×10^{-4}	1.33

Variances incurred by the measurements using the zero-drift IRT were factors of 0.46 and 0.71 smaller than the variances incurred by the precision IRT, for the zero-and steady-state temperature measurements, respectively. The high excess kurtosis values demonstrated that extremal values within the measurements explained much of the variances calculated whilst the IRTs were blanked and, less so, when sighted at the 700 °C blackbody. The ratio of excess kurtosis values indicated that the precision IRT was subject to more extreme variations from its mean value than the zero-drift IRT.

The similarity of variance and excess kurtosis values evaluated for the IRTs measuring 700 °C steady-state temperature, indicated that the two configurations would work similarly well for the measurement of higher temperature sources. The zero-drift IRT did still achieve lower variance than the precision IRT but the difference between the IRTs was not as marked with these higher intensity signals. The zero-drift IRT would be expected to afford lower minimum resolvable temperature measurement, owing to the lower noise and offset, whilst achieving a similar maximum temperature measurable to the precision IRT.

From the zero-signal measurement variances, the noise equivalent power (NEP) within each configuration was calculated to be 1.63×10^{-11} W/$\sqrt{Hz}$ and 2.43×10^{-11} W/$\sqrt{Hz}$ for the zero-drift IRT and precision IRT, respectively. These NEP figures provide an assessment of the minimum detectable power within each IRT, combining both the photodiode and amplifier circuitry.

3.3. Characterisation of IRTs against a Blackbody Source at Low Temperatures

Mean output voltages for the zero-drift IRT and precision IRT, as functions of blackbody source temperature, are presented in Figure 3a. The standard deviations of the measurements are represented by error bars.

The zero-drift IRT measured 0.35 mV at 20 °C, increasing to 2.1 mV at 70 °C, whilst the precision IRT increased from −0.072 mV to 1.4 mV over the same temperature range. Whilst both IRTs demonstrated the ability to resolve changes of target temperature over this measurement range, the standard deviation of the zero-drift IRT measurement was clearly lower across the full measurement range. The mean measurement tolerance of the zero-drift IRT output voltage over the temperature range measured was calculated to be 5.8%, whilst that of the precision IRT was 21.4%. These results reflect what was observed in Figure 2 and Table 1; the zero-drift IRT provides a more sensitive and less noisy measurement.

Whilst the output voltage measurement variation reflected the capability of the thermometers to measure a stable signal, the overall measurement uncertainty included the capability to interpolate the temperature between known calibration points and the calibration tolerance of the equipment used. A more complete estimate of overall measurement uncertainties between 40 °C and 70 °C is presented in the following tables and analysis, thereby affording a more comprehensive understanding of the measurement tolerance of the two versions of the IRTs.

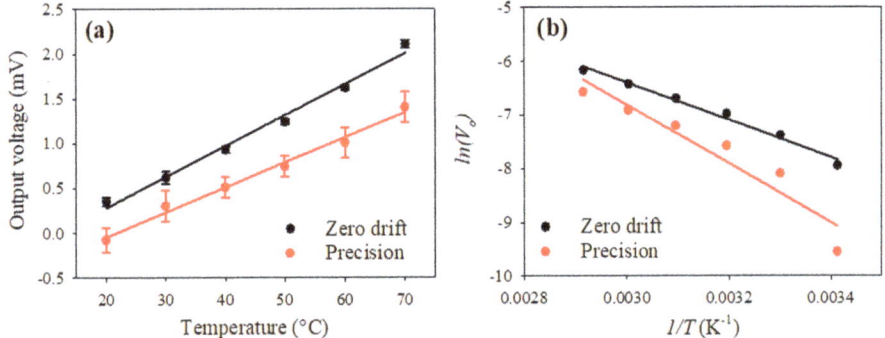

Figure 3. (a) Output voltage as a function of blackbody temperature for zero-drift and precision IRTs. (b) $ln(V_o)$ as a function of $1/T$ for zero-drift and precision IRTs.

The uncertainty budget for single measurements has been estimated, based upon the individual contributions from the equipment used in calibration and measurement and from the measurements acquired using each IRT. The individual uncertainty contributions considered in the estimate, which arose from the calibration equipment and measurements, are presented in Tables 2 and 3, respectively expressed in terms of absolute temperature and percentage of voltage.

Table 2. Contributions of calibration uncertainties to overall measurement uncertainty ($k = 2$).

Source Temperature °C	Calibration RTD Uncertainty K	Calibration Blackbody Uncertainty K	Calibration Instrument Uncertainty K	Digital Multimeter Uncertainty %
40	0.028	0.50	0.16	0.0085
70	0.003	0.45	0.24	0.0085

Table 3. Contributions of measurement uncertainties to overall measurement uncertainty ($k = 2$).

Source Temperature °C	Interpolation Error of Thermometer Mean Measurement K		Infrared Thermometer Voltage Measurement Variability K	
	Zero-Drift IRT	Precision IRT	Zero-Drift IRT	Precision IRT
40	2.3	6.9	1.0	9.5
70	2.7	4.9	0.9	3.5

The largest individual contributions to the uncertainty arose from the stochastic variability in the IRT voltage measurements and the error arising from the interpolation used to convert between the voltage and temperature. The uncertainties that arose from the calibration equipment can be neglected without detriment, therefore we can simplify the uncertainty estimation by evaluating only the variability of the measurements and interpolation errors, presented in Table 3. The overall uncertainties of the IRT measurements, on this basis, varied from 2.5 K at 40 °C to 2.9 K at 70 °C for the zero-drift IRT and 11.7 K at 40 °C and 6.0 K at 70 °C for the precision IRT.

To assess the linearity of the two IRTs, Figure 3b shows $ln(V_o)$ as a function of $1/T$. The zero-drift IRT adhered to the linear relationship expected between $ln(V_o)$ and $1/T$ better than the precision IRT, highlighting an additional benefit of the zero-drift IRT; it eases the calibration procedure. The limiting effective wavelengths for the IRTs were calculated to be 4.12 μm and 3.71 μm for the zero-drift and precision IRTs, respectively.

Practical calibrations of the IRTs were used to determine interpolation curves. There was good agreement between the measured and modelled output voltages for the zero-drift IRT at each temperature within the range, with only the lowest temperature having a deviation that exceeded 10% of the measured voltage. The precision IRT afforded reasonable agreement, with larger deviations between modelled and measured output voltages than recorded for the zero-drift IRT. The difference between modelled and measured voltages for the precision IRT was 47.5% at 20 °C, reducing to circa 30% for temperatures above 40 °C. RMS noise was calculated from the calibrations for each IRT. The RMS noise of the zero-drift IRT was evaluated to be 3.5 K at 20 °C, reducing to less than 1.0 K for temperatures above 40 °C. The RMS noise of the precision IRT was evaluated to be 14.7 K at 20 °C, reducing to 5.0 K for temperatures above 40 °C.

The results have demonstrated that the use of a chopper stabilised op-amp within the TIA circuit of a zero-drift IRT improves the quality of the temperature measurement for target objects at ambient room temperature and above. The zero-drift IRT incurred lower RMS noise compared to an equivalent precision op-amp based IRT. The zero-drift IRT performance could be improved further by deploying a higher value of transimpedance resistor within the feedback loop to achieve higher 'gain'. This would enable lower magnitude photocurrents to be amplified to measurable output voltages. In addition, further investigation of different chopper stabilised op-amp variants could be performed to establish the lower noise performance possible for the zero-drift IRT. By combining this with further instrumentation optimisation, such as additional filtering and averaging, a zero-drift IRT could be developed for measuring temperatures from artefacts such as Lithium ion cells, warm fluids or body temperature. Such an IRT should offer greater stability and be more robust against ambient temperature variations compared with an IRT that used a precision op-amp and could be achieved without photodetector cooling or optical chopping.

4. Conclusions

A fibreoptic coupled zero-drift IRT comprising a chopper stabilised op-amp was shown to offer improved performance compared to a precision IRT containing a precision op-amp. The zero-drift IRT was shown to offer improved linearity for the measurement of lower target temperatures compared to the precision IRT. This offered significant advantages over the precision IRT for the measurement of such temperatures; it eased the IRT calibration procedure and improved the tolerance of the temperature measurement.

When blanked by a cover at ambient temperature, the zero-drift IRT achieved approximately 75% reduction in offset voltage, 50% reduction in the output voltage noise and less susceptibility to perturbation by external sources. This enabled the measurement of lower target temperatures with lower IRT noise; the calibrated RMS noise for the measurement of a 40 °C blackbody source temperature reduced from 5 K to 1 K with thee use of the zero-drift IRT.

The temperature measurement uncertainties of the IRTs between 40 °C and 70 °C, over which range the low-temperature calibrations were compared, were evaluated to vary between 2.5 K and 2.9 K for the zero-drift IRT and 11.7 K and 6.0 K for the precision IRT. The mean uncertainties across the full range of temperature measurements demonstrated that the zero-drift IRT achieved lower measurement uncertainty than the precision IRT.

Author Contributions: Conceptualization, A.D.H., M.J.H. and J.R.W.; methodology, A.D.H. and M.J.H.; validation, A.D.H.; formal analysis, A.D.H. and M.J.H.; investigation, A.D.H.; resources, J.R.W.; data curation, A.D.H.; writing—original draft preparation, A.D.H. and M.J.H.; writing—review and editing, A.D.H., M.J.H. and J.R.W.; supervision, J.R.W.; project administration, A.D.H.;funding acquisition, J.R.W. All authors have read and agreed to the published version of the manuscript.

Funding: This research was funded by the UK Engineering and Physical Sciences Research Council under doctoral training grant scholarship numbers EP/K503149/1 and EP/L505055/1 and grant number EP/M009106/1 Optimised Manufacturing Through Unique Innovations in Quantitative Thermal Imaging and the APC was funded by EPSRC/UKRI and administered by The University of Sheffield.

Conflicts of Interest: The authors declare no conflicts of interest. The funders had no role in the design of the study; in the collection, analyses or interpretation of data; in the writing of the manuscript, and in the decision to publish the results.

References

1. Childs, P.R.N.; Greenwood, J.R.; Long, C.A. Review of temperature measurement. *Rev. Sci. Instrum.* **2000**, *71*, 2959–2978. [CrossRef]
2. Demling, A.; Ousley, D.; Stelley, S. Best practices for deploying thermocouple instruments. *AIP Conf. Proc.* **2013**, *1552*, 601–606. [CrossRef]
3. Martyniuk, P.; Rogalski, A. Terahertz detectors and focal plane arrays. *Opto-Electron. Rev.* **2013**, *21*, 239–257. [CrossRef]
4. Heeley, A.D.; Hobbs, M.J.; Laalej, H.; Willmott, J.R. Miniature uncooled and unchopped fiber optic infrared thermometer for application to cutting tool temperature measurement. *Sensors* **2018**, *18*, 3188. [CrossRef]
5. Moisello, E.; Vaiana, M.; Castagna, M.E.; Bruno, G.; Malcovati, P.; Bonizzoni, E. An integrated thermopile-based sensor with a chopper-stabilized interface circuit for presence detection. *Sensors* **2019**, *19*, 3999. [CrossRef]
6. Makai, J.P.; Makai, T. Enhancement of the low level detection limit of radiometric quality photovoltaic and photoconductive detectors. *Metrologia* **2005**, *42*, 266–270. [CrossRef]
7. Dixon, J. Radiation thermometry. *J. Phys. E Sci. Instrum.* **1988**, *21*, 425–436. [CrossRef]
8. Srivastav, V.; Sharma, R.K.; Bhan, R.K.; Dhar, V.; Venkataraman, V. Exploring novel methods to achieve sensitivity limits for high operating temperature infrared detectors. *Infrared Phys. Technol.* **2013**, *61*, 290–298. [CrossRef]
9. Texas Instruments Inc. *AN-1803 Design Considerations for a Transimpedance Amplifier*; Texas Instruments Appl. Rep. SNOA515A; Texas Instruments: Dallas, TX, USA, May 2013; pp. 1–6. Available online: http://www.ti.com/lit/an/snoa515a/snoa515a.pdf (accessed on 24 February 2016).
10. Orozco, L. *Optimizing Precision Photodiode Sensor Circuit Design*; Analog Devices Technical Article MS-2624; Analog Devices: Norwood, MA, USA, 2014; pp. 1–5. Available online: https://www.analog.com/media/en/technical-documentation/tech-articles/Optimizing-Precision-Photodiode-Sensor-Circuit-Design-MS-2624.pdf (accessed on 10 April 2017).
11. Eppeldauer, G. Chopped radiation measurements with large area Si photodiodes. *J. NIST* **1998**, *103*, 153–162. [CrossRef]
12. Portelli, A.J.; Nasuto, S.J. Design and development of non-contact bio-potential electrodes for pervasive health monitoring applications. *Biosensors* **2017**, *7*, 2. [CrossRef]
13. Wang, Y.; Wunderlich, R.; Heinen, S. A micropower analogue front end for wireless ECG system A micropower analogue front end for wireless ECG system. *Int. J. Electron. Lett.* **2018**, *6*, 36–47. [CrossRef]
14. Kim, J.; Kim, H.; Han, K.; You, D.; Heo, H.; Kwon, Y.; Cho, D.D.; Ko, H. Low-noise chopper-stabilized multi-path operational amplifier with nested miller compensation for high-precision sensors. *Appl. Sci.* **2020**, *10*, 281. [CrossRef]
15. Han, K.; Kim, H.; Kim, J.; You, D.; Heo, H.; Kwon, Y.; Lee, J.; Ko, H. A 24.88 nv/√hz wheatstone bridge readout integrated circuit with chopper-stabilized multipath operational amplifier. *Appl. Sci.* **2020**, *10*, 399. [CrossRef]
16. Butti, F.; Bruschi, P.; Dei, M.; Piotto, M. A compact instrumentation amplifier for MEMS thermal sensor interfacing. *Analog Integr. Circuits Signal. Process.* **2012**, *72*, 585–594. [CrossRef]
17. Makai, J.P.; Makai, J.J. Current-to-voltage converter for linearity correction of low shunt resistance photovoltaic detectors. *Rev. Sci. Instrum.* **1996**, *67*, 2381–2386. [CrossRef]
18. Aleksandrov, S.E.; Gavrilov, G.A.; Sotnikova, G.Y. Effect of low-frequency noise on the threshold sensitivity of middle-IR photodetectors in a broad frequency range. *Tech. Phys. Lett.* **2014**, *40*, 704–707. [CrossRef]
19. Horowitz, P.; Hill, W. Table 5.2 Representative Precision Op-amps. In *The Art of Electronics*, 3th ed.; Cambridge University Press: Cambridge, UK, 2015; p. 302.

20. Horowitz, P.; Hill, W. Table 5.6 Chopper and Auto-zero Op-amps. In *The Art of Electronics*, 3th ed.; Cambridge University Press: Cambridge, UK, 2015; p. 335.
21. Linear Technologies. *LT1012A/LT1012-Picoamp Input Current, Microvolt Offset, Low Noise Operational Amplifier Datasheet*; LW/TP 1202 1K Rev., B.; Linear Technologies: Milpitas, CA, USA, 1991; pp. 1–20. Available online: https://www.analog.com/media/en/technical-documentation/data-sheets/LT1012.pdf (accessed on 12 June 2017).
22. Linear Technologies. *LTC2050/LTC2050HV Zero Drift Operational Amplifiers Datasheet*; LT0817 Rev., D.; Linear Technologies: Milpitas, CA, USA, 1999; pp. 1–18. Available online: https://www.analog.com/media/en/technical-documentation/data-sheets/LTC2050-2050HV.pdf (accessed on 18 December 2017).
23. Butti, F.; Piotto, M.; Bruschi, P. A chopper instrumentation amplifier with input resistance boosting by means of synchronous dynamic element matching. *IEEE Trans. Circuits Syst. Regul. Pap.* **2017**, *64*, 753–764. [CrossRef]
24. Lee, S.; Shin, Y.; Kumar, A.; Kim, K.; Lee, H. Two-Wired Active Spring-Loaded Dry Electrodes for EEG measurements. *Sensors* **2019**, *19*, 4572. [CrossRef]
25. Nebhen, J.; Meillère, S.; Masmoudi, M.; Seguin, J.L.; Barthelemy, H.; Aguir, K. A 250 μw 0.194 nV/rtHz chopper-stabilized instrumentation amplifier for MEMS gas sensor. In Proceedings of the 7th International Conference on Design & Technology of Integrated Systems in Nanoscale Era, Gammarth, Tunisia, 16–18 May 2012; pp. 1–5. [CrossRef]
26. Zhao, J.; Zhang, S.; Chen, S. A chopper stabilized pre-amplifier for magnetic sensor. In Proceedings of the 2012 International Conference on Industrial Control and Electronics Engineering, Xi'an, China, 23–25 August 2012; pp. 501–504. [CrossRef]
27. Sparrow, E.M.; Albers, L.U.; Eckert, E.R.G. Thermal radiation characteristics of cylindrical enclosures. *J. Heat Transf.* **1962**, *84*, 73–81. [CrossRef]
28. Vollmer, J. Study of the effective thermal emittance of cylindrical cavities. *J. Opt. Soc. Am.* **1957**, *47*, 926–932. [CrossRef]
29. Saunders, P. General interpolation equations for the calibration of radiation thermometers. *Metrologia* **1997**, *34*, 201–210. [CrossRef]
30. Sakuma, F.; Hattori, S. Establishing a practical temperature standard by using a narrow-band radiation thermometer with a silicon detector. In *Temperature: Its Measurement and Control in Science and Industry, Proceedings of the Sixth International Temperature Symposium, Washington, DC, USA, 15–18 March 1982*; American Institute of Physics: College Park, MD, USA, 1982; Volume 5, pp. 421–427.
31. Pfeifer, H. Industrial Furnaces-Status and Research Challenges. *Energy Procedia* **2017**, *120*, 28–40. [CrossRef]

 © 2020 by the authors. Licensee MDPI, Basel, Switzerland. This article is an open access article distributed under the terms and conditions of the Creative Commons Attribution (CC BY) license (http://creativecommons.org/licenses/by/4.0/).

Article

Study of Image Classification Accuracy with Fourier Ptychography

Hongbo Zhang [1,*], Yaping Zhang [2,*], Lin Wang [3], Zhijuan Hu [4], Wenjing Zhou [5], Peter W. M. Tsang [6], Deng Cao [1] and Ting-Chung Poon [3]

1. College of Engineering, Science, Technology and Agriculture (CESTA), Central State University, Wilberforce, OH 45384, USA; dcao@centralstate.edu
2. Yunnan Provincial Key Laboratory of Modern Information Optics, Kunming University of Science and Technology, Kunming 650500, China
3. Bradley Department of Electrical and Computer Engineering, Virginia Tech, Blacksburg, VA 24060, USA; linwang@vt.edu (L.W.); tcpoon@vt.edu (T.-C.P.)
4. College of Mathematics and Science, Shanghai Normal University, 100 Guilin Road, Shanghai 200234, China; huzhijuan@shnu.edu.cn
5. Department of Precision Mechanical Engineering, Shanghai University, Shanghai 200444, China; lazybee@shu.eud.cn
6. Department of Electrical Engineering, City University of Hong Kong, 83 Tat Chee Avenue, Kowloon, Hong Kong 00852, China; eewmtsan@cityu.edu.hk
* Correspondence: hzhang@centralstate.edu (H.Z.); yaping.zhang@gmail.com (Y.Z.)

Abstract: In this research, the accuracy of image classification with Fourier Ptychography Microscopy (FPM) has been systematically investigated. Multiple linear regression shows a strong linear relationship between the results of image classification accuracy and image visual appearance quality based on PSNR and SSIM with multiple training datasets including MINST, Fashion MNIST, Cifar, Caltech 101, and customized training datasets. It is, therefore, feasible to predict the image classification accuracy only based on PSNR and SSIM. It is also found that the image classification accuracy of FPM reconstructed with higher resolution images is significantly different from using the lower resolution images under the lower numerical aperture (NA) condition. The difference is yet less pronounced under the higher NA condition.

Keywords: fourier ptychography; image classification; deep learning; neural network

1. Introduction

Fourier Ptychography Microscopy (FPM) is a computational microscopy imaging technique potentially able to achieve a wide field of view and high-resolution imaging. It involves the use of a sequence of LEDs (LED array), which illuminates sequentially onto the target. Based on the sequential illumination, iterated sampling methods in the frequency domain are used for recovering higher resolution images [1–3]. Different types of Fourier Ptychography techniques have been proposed in the past for improving imaging resolutions [1–7]. Among them, multiplexed coded illumination techniques, laser-based implementations, aperture scanning Fourier Ptychography, camera scanning Fourier Ptychography, multi-camera approach, single-shot Fourier Ptychography, speckle illumination, X-ray Fourier Ptychography, and diffuser modulation have achieved successes [1]. More recent research has specifically addressed the Fourier Ptychography imaging problems such as the brightfield, phase, darkfield, reflective, multi-slice, and fluorescence imaging [3–5].

Driven by the significant interests in deep learning, a few different methods have been developed to solve the ill-posed Fourier Ptychography imaging problems. Notably, a convolutional neural network has been successfully applied for solving the Fourier Ptychography imaging problem [7]. The U-Net type structure-based generative adversarial network (GAN) was used to utilize fewer samples (26 images versus a few hundred images) for obtaining similar quality image reconstruction [8]. More promisingly, the U-Net plus GAN structure is proven able to reconstruct the high-resolution images using only a smaller amount of lower resolution image input data [9]. An interpretable deep learning approach has shown its effectiveness for the imaging of scattering materials [9,10].

Until recently, the majority of Fourier Ptychography imaging studies have been primarily focusing on achieving high-resolution imaging based on the criteria of human visual perception satisfaction, yet a significant question left to answer is to understand the impact of such a technique on the broad spectrum of the downstream image processing related visual tasks such as classification, segmentation, and object detection. The ability to answer these questions is critical for applications such as industrial robotics, medical imaging, and industrial automation [10,11].

It is well known that deep learning is time-consuming, thus imposing significant computational challenges. However, a significant research effort has been made where the recent breakthrough has shown that the training time of Imagenet has been reduced from days to hours, even further to minutes [12]. All these breakthroughs, however, come with a price. They all require expensive GPU devices for training the deep learning neural network. The expenses of the costly hardware can be well reflected through the price of the GPU card, for example, an NVIDIA P-100 GPU card costs about USD 6000 with the current market [12]. Training of all possible combinations of the different quality of FPM reconstructions for obtaining image classification accuracy is feasible but time-consuming and costly. As such, it is desirable to formulate the relationships between the FPM reconstructed image quality and the image classification accuracy in order to avoid training all the different combinations of the data and still able to identify the optimal parameters used for FPM reconstruction to achieve the best performance of the image classification accuracy [13–20].

With this research, we propose to use the Fourier Ptychography technique to reconstruct higher (FPM-based) and lower resolution (without FPM) images for image classification. Following the FPM reconstruction, a deep convolution neural network is constructed for evaluation of the image recognition accuracy for both higher and lower resolution images. A multiple linear regression model is used to regress the relationship between independent variables including peak signal to noise ratio (PSNR), the structural similarity index (SSIM), and the dependent variable image classification accuracy. Based on the regression model, it becomes feasible to infer the image classification accuracy directly based on PSNR and SSIM rather than through the intensive and time-consuming deep learning-based image classification training. In comparison to deep learning image classification training, PSNR and SSIM are easier to calculate with lower computational cost, the proposed method then becomes useful for predicting the image classification accuracy of FPM-related visual tasks. Additionally, our research also provides insights into the general effectiveness of image classification accuracy following FPM reconstruction by comparing the FPM reconstruction with or without FPM reconstruction (also known as higher or lower resolution images). In conclusion, our contribution of using PSNR and SSIM to predict image classification accuracy is not only limited to the FPM technique but also a universal approach to other imaging techniques such as digital holography, optical scanning holography, and transport of intensity imaging [21,22].

2. Methodology

2.1. Fourier Ptychography

For the imaging task, given an object complex field $O_0(x, y)$ with spectrum represented by $O_0(k_x, k_y)$, where k_x and k_y denote the spatial frequencies along the x and y-direction. The n-th captured low-resolution raw image intensity is $I_n(x, y)$ (n = 1, 2, 3 ... N) with a spectrum of $G_n(k_x, k_y)$. The coherent transfer function (CTF) of the microscope objective is given as $C(k_x, k_y) = circ(k_r/k_0 NA)$, where $k_r = \sqrt{k_x^2 + k_y^2}$, k_x and k_y represent the spatial frequencies along the x and y directions. denotes a value 1 within a circle of radius r_0 and 0 otherwise, and $r = \sqrt{x^2 + y^2}$. NA is the numerical aperture of the microscope objective and finally k_0 is the wavenumber of the light source.

The FPM reconstruction starts from an initial guess of the spectrum distribution of the nth raw image, $G_n^0(k_x, k_y)$, according to:

$$g_n^0(x,y) = \Im^{-1}\{G_n^0(k_x, k_y)\} = \Im^{-1}\{O_0(k_x - k_{xn}, k_y - k_{yn})C(k_x, k_y)\} \quad (1)$$

where $\Im$ and $\Im^{-1}$ represent Fourier transform and inverse Fourier transform. k_{xn} and k_{yn} indicate the phase-shift caused by the oblique illumination along the x and y directions. To update g_n^0, we replace the amplitude of g_n^0 with $\sqrt{I_n}$ as:

$$g_n^1(x,y) = \sqrt{I_n}\frac{g_n^0}{|g_n^0|} \quad (2)$$

The superscript indices 0 and 1 in $g_n(x,y)$ represent the prior and post updated low-resolution image, respectively. The updated spectrum is $G_n^1(k_x, k_y) = \Im\{g_n^1(x,y)\}$, as such the updated spectrum $O_0^1(k_x, k_y)$ is:

$$O_0^1(k_x - k_{xn}, k_y - k_{xn}) = O_0(k_x - k_{xn}, k_y - k_{yn})[1 - C(k_x, k_y)] + G_n^1(k_x, k_y) \quad (3)$$

Repeating Equations (1) to (3) by using I_n from $n = 1$ to $n = N$ in one iteration, following convergence, the error function E_k becomes:

$$E_k = \sum_{n=1}^{N}\sum_{x,y}\left\{\left|g_{k,n}^1(x,y)\right| - \sqrt{I_n(x,y)}\right\}^2, \quad (4)$$

The second summation indicates the pixel-by-pixel summation for every single image. In $g_{k,n}^0$, the superscript indices 0 and k represent the complex distribution of the n-th low-resolution image before the k-th iteration is completed. Following the k-th iteration, the high-frequency component-maintained spectrum $O_k(k_x, k_y)$ is recovered. The intensity distribution of it is:

$$I_k = |O_k(x,y)|^2 = \left|\Im^{-1}\{O_k(k_x, k_y)\}\right|^2 \quad (5)$$

$O_k(k_x, k_y)$ of Equation (5) indicates the desired spectrum after the k-th iteration, correspondingly, the $O_0(k_x, k_y)$ indicates the original spectrum prior to the iteration of the update, the same as to I_k and $O_k(x,y)$. Figure 1 shows the sequence of the FPM algorithm.

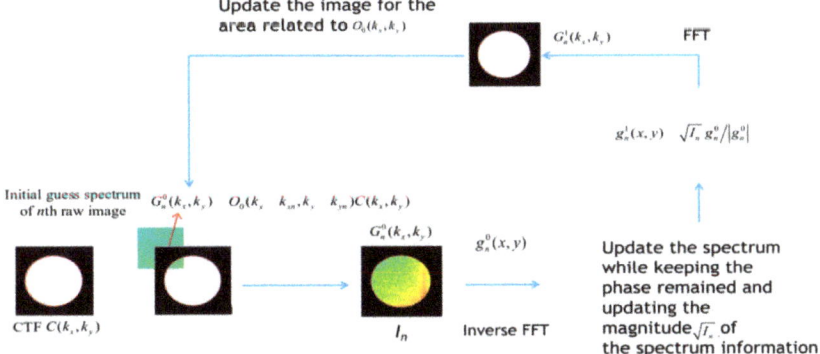

Figure 1. The iteration process of FPM. In the iteration, the sampling rate of the initial guess of the high-resolution object is higher than the collected low-resolution images. Through the iterative process, the reconstructed image is increased in spatial resolution.

In simulation, we choose to use a 630 nm red laser. Camera CCD resolution is 2.76 µm. The distance between the LED array and the sample is 90 mm. The gap between LEDs is 4 mm. The Fourier Ptychography image reconstruction process starts from the lower resolution direct images. In our study, we used 225 slower resolution images (15 × 15 LED array). Based on the 225 lower resolution images, we continued to perform spectrum sampling in the frequency domain. Based on Equation (2), the higher resolution image spectrum was updated until the converging condition specified was reached as shown in Equation (3). For the purpose of comparison between the FPM reconstructed image (the higher resolution) and the lower resolution image (the raw lower resolution image, without going through the FPM reconstruction process), the middle LED of the 15 by 15 LED array was chosen to be used. The system numerical aperture (NA) chosen here was varied from 0.05 to 0.5.

2.2. Image Classification

For the evaluation of the accuracy of image classification, a deep convolution neural network was used. We used six different neural networks for systematic image classification. The first deep convolution neural network was designed to train the MNIST dataset. The architecture of the convolution neural network is shown in Figure 2. In this convolution neural network, the following structure was used. Given the input size of 28 by 28 images, a 3 by 3 convolution is performed. Following the 3 by 3 convolution, a feature map of size 26 by 26 was obtained. There was a total of 16 such feature maps obtained. A 2 by 2 max-pooling was consequently performed. It produced a 13 by 13 feature map, where in total 16 such feature maps were obtained. A 3 by 3 convolution was further processed. Therefore, this lead to 32 10 by 10 feature maps. Continuously, a 2 by 2 max-pooling was conducted, yielding 32 feature maps at the size of 5 by 5. It follows that the 3 by 3 convolution operation yielded a 2 by 2 feature map with a total of 64 such feature maps. The fully connected operation was also conducted yielding a fully connected layer at the size of 10. Finally, softmax was performed for image classification.

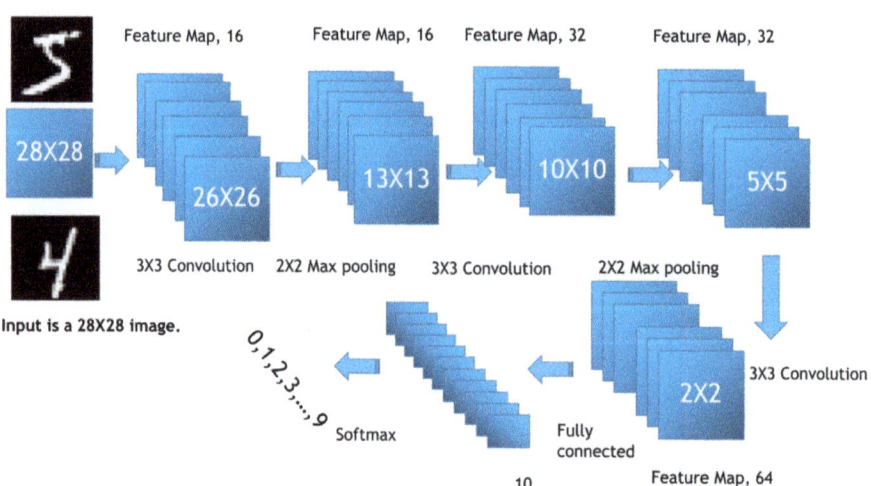

Figure 2. Convolution neural network architecture for image classification using MNIST data.

Similarly, the same network architecture was also used for the Fashion-MNIST dataset training as shown in Figure 3. The Fashion-MNIST dataset is built following the principle of MNIST dataset. It has an identical format to the MNIST dataset in terms of image size, but has a more complex geometry including 11 classes of t-shirt, trouser, dress, coat, sandal, shirt, sneaker, bag, and ankle boot [14]. It is known that Fashion-MNIST is a more challenging data benchmark for image classification tasks [15].

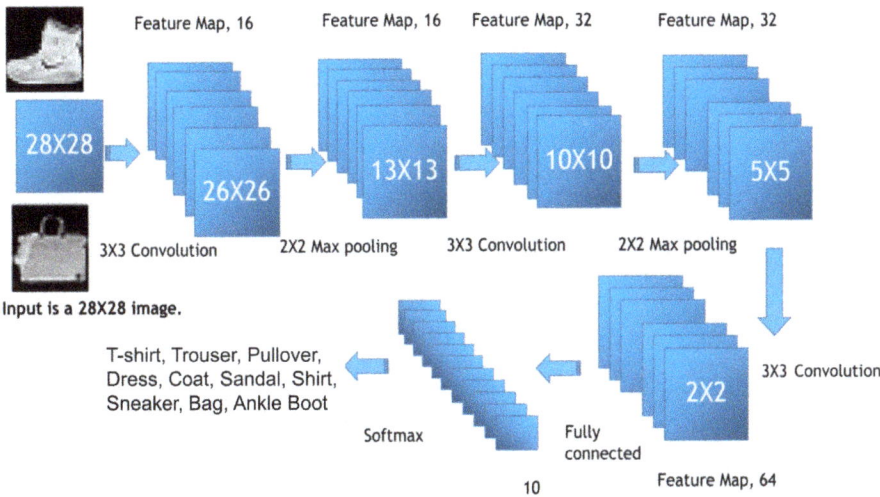

Figure 3. Convolution neural network architecture for image classification using Fashion-MNIST data.

It is worthwhile to note that the purpose of our research was not to achieve the highest image classification accuracy, but rather, we wanted to compare the image classification accuracy between an FPM reconstructed image (higher resolution) versus the lower resolution image (raw lower resolution image collected). We also want to discover the relationship between PSNR, SSIM, and image classification accuracy. As such, we only randomly chose part of the entire 60,000 images. Through our experiment, we discovered that the use of a partial set of images is able to achieve reasonable training accuracy while preventing overfitting. Thus, in this research, we chose to use 2560 MNIST images for performing training. By doing this, there are two advantages, first, the use of fewer images enabled us to do more rapid training. For example, training 50,000 FPM-processed images requires over 30 h on a moderate performance laptop. Yet, training of 2560 images requires nearly 20 times less training time. This allows us to do multiple trainings given different NAs within a reasonable time frame, thus were are able to build the regression model between PSNR, SSIM, and classification accuracy. However, the principles of the method proposed in this research were able to be universally applied to an arbitrary number of images for training and evaluation.

Similarly, for Fashion-MNIST, 2560 FPM processed images were used for training. A total of 512 FPM-processed images were used for the evaluation of the classification accuracy. Stochastic gradient descent with momentum was used as the optimizer for both MNIST and Fashion-MNIST training. Compared to batch stochastic gradient descent, minibatch stochastic gradient descent is more capable of reducing the training error, as such, it is also adopted to reduce the training error to the smallest. The training learning rate was chosen as 0.0001. For all the convolution layers, a padding of 1 was used. Eighty epochs were iterated before the training stopped.

The CIFAR dataset was also used for both training and evaluation. The CIFAR dataset includes 10 classes of airplane, automobile, bird, cat, deer, dog, frog, horse, ship, and truck. Differently, images from the CIFAR dataset have three channels (RGB), whereas both MNIST and Fashion-MNIST only have a single channel. Among them, similar to MNIST and Fashion-MNIST, 2560 FPM-processed images were used for training, and 512 FPM-processed images were used for evaluation of the classification accuracy. The residual type of neural network structure was adopted for training and evaluation purposes. The residual network structure is known to enable deeper layers of a network without the vanishing gradient problem [16]. It suggests that even with deeper layers, the training is still able to converge within a reasonable amount of training time. The residual network is also able to perform well on image feature maps fusion, taking advantage of its unique long-short memory mechanisms, and thus achieving better image classification accuracy [16]. The learning rate was chosen as 0.1. In total, 80 epochs during training were conducted. The residual network structure is shown in Figure 4. In order to reduce the likelihood of over-fitting, image augmentation is conducted including image translation and reflection. Cropping is performed prior to the start of training [17].

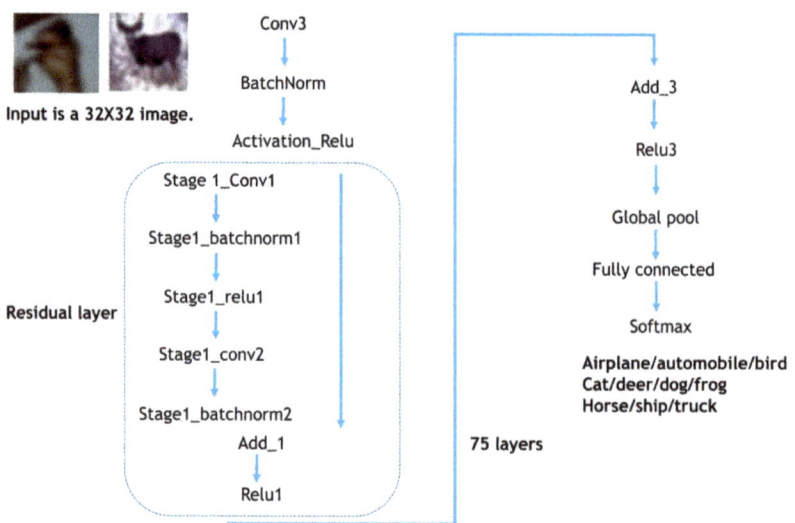

Figure 4. Residual convolution neural network architecture for image classification using the CIFAR dataset. The line (on the right side) in the residual layer represents a skipped connection.

The Caltech 101 dataset ass also used for training. Within Caltech 101, there are 101 classes including plane, chair, soccer, brain, and others. Each class has about 60 images. In total, 25 categories were used for training and evaluation. In contrast to MNIST, Fashion-MNIST and the CIFAR image set have limited resolutions of 28 by 28 and 32 by 32. Caltech 101 image resolution is much higher, mostly with a size of 300 by 200. The use of large images can better ensure the FPM image reconstruction accuracy. The FPM reconstruction method involves iteratively sampling the overlap region of the lower resolution images in the frequency domain. As such, a larger input can help to improve the reconstruction accuracy [1]. With the inclusion of larger input images, the input dataset becomes richer. With the richer input dataset, it is also helpful to build a more diversified relationship model between image classification, PSNR, and SSIM. Similar to the training performed against the CIFAR dataset, the deep residual network, resnet50, was used for the image classification task shown in Figure 5. The neural network had 177 layers connected by the residual blocks among the inter layers. Before the training started, color and cropping-based image augmentation was performed to reduce the likelihood of image overfitting [17,18]. Note that the linear activation function (Relu) was used in the residual network. The function was able to train a deeper neural network without the vanishing gradient problem because the activation function had the advantages of both linear and nonlinear transformations of the input. The batch normalization reduced the value variation for each layer, thus increasing the stability of the deep network training and reducing the needed epochs for achieving the ideal classification accuracy.

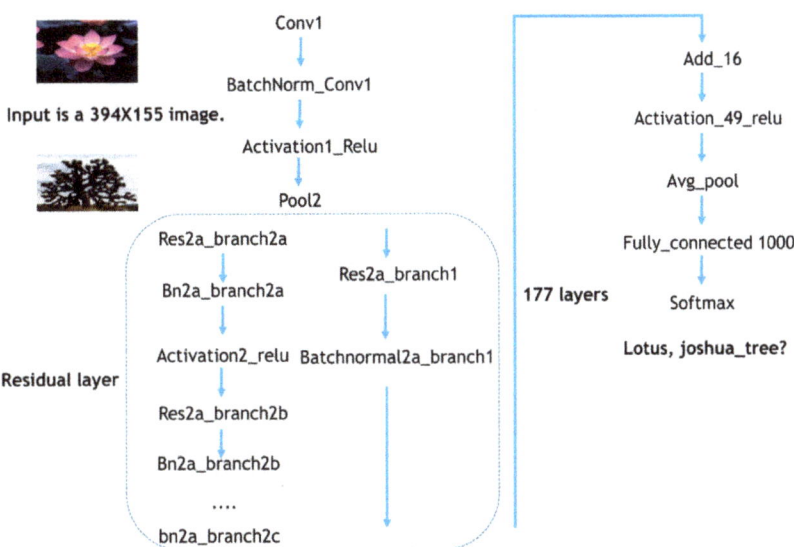

Figure 5. Residual convolution neural network architecture for image classification training using the Caltech 101 dataset. Res2s_branch2a represents the second stage and the second branch.

The use of customized data for testing the image classification accuracy with and without FPM reconstruction was also conducted. For this, two customized training data were included, namely the flowers and apple pathology datasets. The flowers dataset included 4242 images with sub-categories of daisy, dandelion, flowers, rose, sunflower, and tulip. The number of images per category were balanced. The balanced apple pathology data had 3200 images. Blackrot, cedar rust, scab, and healthy apples were included. For the flowers dataset, SqueezeNet was used for the classification of different types of flowers shown in Figure 6. The SqueezeNet consists of repeated Squeeze and expansion neural network modules. The Squeeze and fire modules employ a 1 by 1 filter. The direct benefit of using a 1 by 1 convolution filter is that the network requires fewer parameters and more memory efficient. The subsequent downsampling (e.g., pool10) enables the larger activation map, thus is beneficial for maximizing the classification accuracy.

For the apple pathology dataset, a Google inception-based network was used for the classification of the apple diseases shown in Figure 7. Google inception employs the inception layer as the backbone of the network. In total, nine repeated inception modules were used. Within the inception modules, different sizes of filters were used. Among them, 1 by 1, 3 by 3, and 5 by 5 filters were used within the inception module. By doing this, it can save weight parameters of the deep network. Additionally, by dividing the sequential filters into four branches within the inception layer, it further reduced the numbers of parameters, thus enabling a further increase in the computational efficiency.

Through training, a fine-tuned approach is utilized. The beginning three layers of the network are frozen and only the last two layers are trained. The data are augmented through reflection along one side of the image, followed by random translation and scaling along both sides of the image. A minimum batch size of 5 was used per batch throughout the training process. Random shuffling of the data was performed to reduce the likelihood of over-fitting. Twenty percent of the data was used for validation of the training accuracy. For achieving good training accuracy, a small initial learning rate of 0.0003 was used. In total, 6 epochs of training was found to be sufficient for achieving a converged training accuracy.

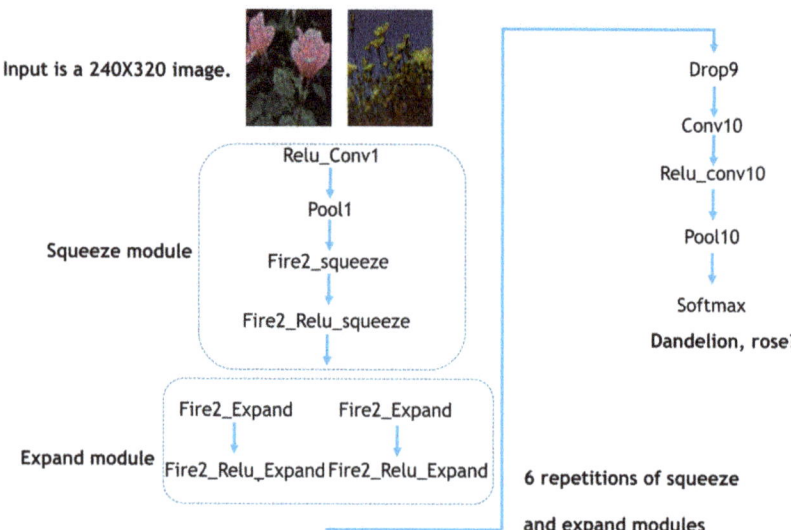

Figure 6. SqueezeNet convolution neural network architecture for image classification training using the flowers dataset. Fire2_squeeze means the second stage squeeze module.

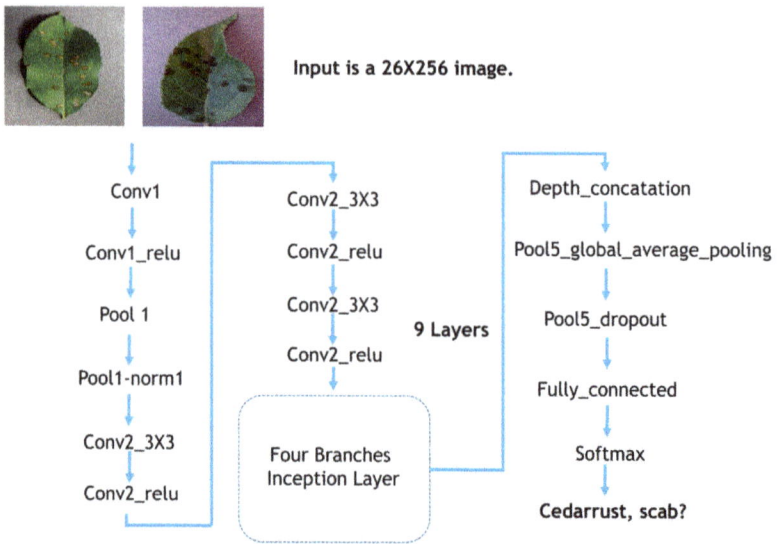

Figure 7. Google inception convolution neural network architecture for image classification training using the apple pathology dataset. Conv2_3 × 3 represents the second stage, 3 × 3 convolution filter.

Following the training of the MNIST, Fashion-MNIST, CIFAR, Caltech 101, flowers, and apple pathology datasets, the image classification accuracy was obtained for the FPM reconstructed image, lower resolution image, as well as the ground truth image. Furthermore, PSNR and SSIM are also calculated for the FPM reconstructed image, lower resolution image, as well as ground truth image. The calculation of PSNR and SSIM values uses the ground truth image as the reference. Multiple linear regression is also performed for the PSNR, SSIM (independent variables), and image classification accuracy (dependent variable). A *p*-value of 0.05 is used as the significance threshold of the F-Test performed against the goodness of fit of the regression.

3. Results

The results of Fourier Ptychography are shown in Figure 8, Figure 9, Figure 10 for different datasets. Figure 8 shows the results using MNIST and Fashion-MNIST.

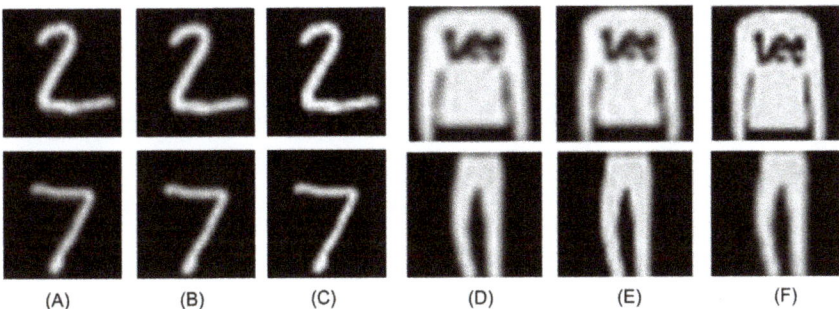

Figure 8. (**A**) Lower resolution MNIST image; (**B**) FPM reconstructed MNIST image; (**C**) original MNIST image; (**D**) lower resolution Fashion MNIST image; (**E**) FPM reconstructed fashion MNIST image; (**F**) original Fashion MNIST image.

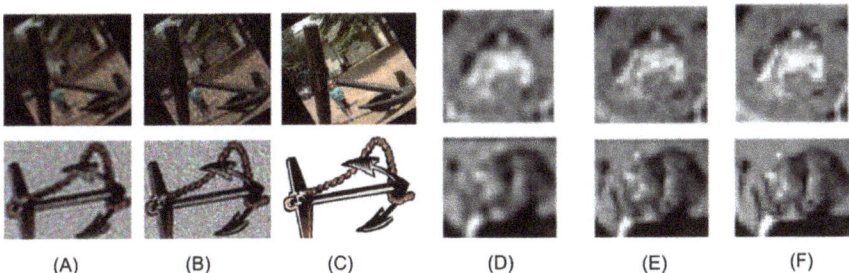

Figure 9. (**A**) Lower resolution CalTech 101 image, (**B**) FPM reconstructed CalTech 101 image (**C**) Original CalTech 101 image (**D**) Lower resolution Fashion CIFAR image (**E**) FPM reconstructed CIFAR image (**F**) original CIFAR image.

Figure 10. (**A**) Lower resolution apple pathology image; (**B**) FPM-reconstructed apple pathology image; (**C**) original apple pathology image; (**D**) lower resolution flowers image; (**E**) FPM-reconstructed flowers image; (**F**) original flowers image.

Lower resolution image, FPM reconstructed image, and original image based on CIFAR and CalTech 101 are shown in Figure 9. Similarly, the lower resolution image, FPM reconstructed image, and original image based on the flowers and apple pathology datasets are shown in Figure 10. Consistently, all four datasets show their effectiveness in the generation of higher resolution images using FPM reconstruction.

The results of PSNR and SSIM, and classification accuracy for CIFAR, CalTech101, MNIST, Fashion-MNIST, flowers, and apple pathology datasets are shown in Table 1, Table 2, Table 3.

Table 1. PSNR, SSIM, and classification accuracy for CIFAR and CalTech101.

Dataset	Numerical Aperture	Reconstruction Method	PSNR	SSIM	Classification Accuracy
CIFAR		Ground Truth	Inf	1	76.61%
	0.5	FPM	14.64	0.79	61.72%
		Lower Resolution	14.39	0.76	58.79%
	0.2	FPM	14.46	0.78	59.38%
		Lower Resolution	13.81	0.61	52.74%
	0.05	FPM	12.48	0.33	42.39%
		Lower Resolution	12.06	0.31	35.16%
CalTech 101		Ground Truth	Inf	1	91.75%
	0.5	FPM	13.37	0.52	73.75%
		Lower Resolution	12.82	0.50	61.50%
	0.2	FPM	13.11	0.53	65.50%
		Lower Resolution	12.44	0.49	36.25%
	0.05	FPM	11.09	0.31	20.00%
		Lower Resolution	10.69	0.38	4.75%

Table 2. PSNR, SSIM, and classification accuracy for MNIST and Fashion-MNIST.

Dataset	Numerical Aperture	Reconstruction Method	PSNR	SSIM	Classification Accuracy
MNIST		Ground Truth	Inf	1	96.87%
	0.5	FPM	27.82	0.84	96.29%
		Lower Resolution	22.11	0.53	95.89%
	0.2	FPM	23.99	0.63	97.66%
		Lower Resolution	16.81	0.33	94.73%
	0.05	FPM	12.12	0.13	89.06%
		Lower Resolution	11.58	0.05	62.50%
Fashion MNIST		Ground Truth	Inf	1	85.35%
	0.5	FPM	26.88	0.88	84.57%
		Lower Resolution	22.69	0.72	83.59%
	0.2	FPM	23.88	0.76	83.20%
		Lower Resolution	17.50	0.47	83.01%
	0.05	FPM	11.96	0.19	78.23%
		Lower Resolution	11.94	0.11	67.58%

The multiple variable linear regression between independent variables of PSNR and SSIM and the dependent variable of classification accuracy was performed. Figure 11 shows the linear relationships between SSIM, PSNR, and classification accuracy. As shown in Figure 11, the linear relationship between image classification accuracy, PSNR, and SSIM is evident.

Table 3. PSNR, SSIM, and classification accuracy for flowers and apple pathology datasets.

Dataset	Numerical Aperture	Reconstruction Method	PSNR	SSIM	Classification Accuracy
Flowers		Ground Truth	Inf	1	83.15%
	0.5	FPM	19.26	0.80	81.25%
		Lower Resolution	18.42	0.68	74.19%
	0.2	FPM	18.66	0.70	76.30%
		Lower Resolution	17.35	0.59	72.14%
	0.05	FPM	16.02	0.41	70.28%
		Lower Resolution	15.33	0.47	49.18%
Apple Pathology		Ground Truth	Inf	1	99.56%
	0.5	FPM	15.44	0.62	98.12%
		Lower Resolution	15.32	0.66	97.32%
	0.2	FPM	15.28	0.66	91.23%
		Lower Resolution	14.98	0.67	89.19%
	0.05	FPM	6.71	0.0082	90.12%
		Lower Resolution	6.64	0.0003	82.35%

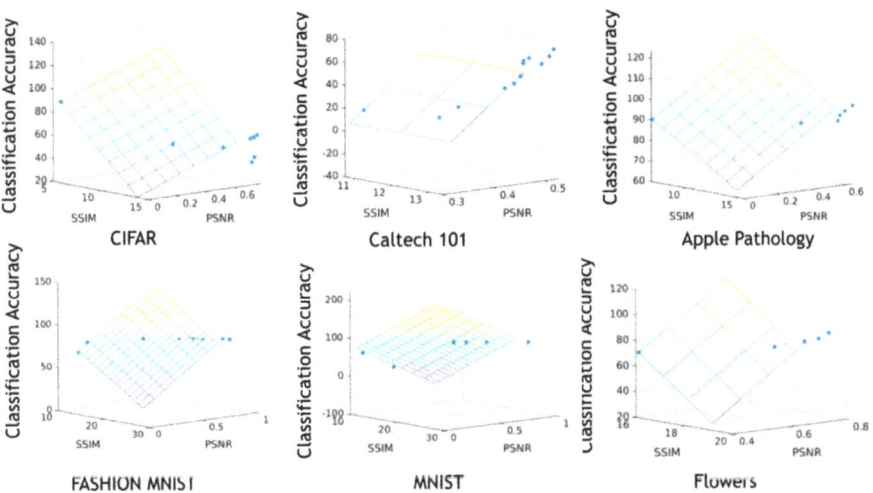

Figure 11. Visualization of data distribution for multiple linear regression between image classification accuracy, PSNR, and SSIM.

The p-values, an indicator of the efficacy of the multiple linear regression, are shown in Table 4. It is also clear that all the F-Test p-values are less than 0.05 showing that a significant linear relationship exists between these variables within the linear regression model.

PSNR and SSIM increased with higher values of numerical aperture. For the purpose of illustrating the increased image classification accuracy due to the increase in PSNR and SSIM, the neural network activation map under different numerical apertures is presented. Figure 12 shows the Google inception convolution neural network activation map and the reconstructed images corresponding to each numerical aperture. The reconstructed image on the top of Figure 12 is the input of the Google inception neural network, and the activation map is the gradient of the features learned from the neural network.

Table 4. F-Test *p*-values for multiple linear regression between image classification accuracy, PSNR, and SSIM for MNIST, Fashion-MNIST, CIFAR, CalTech 101, and the flowers and apple pathology datasets.

Multiple Linear Regression *p*-Value	*p*-Value	Statistical Significance ($p < 0.05$)
MNIST	0.0046	Yes
Fashion-MNIST	3.02×10^{-5}	Yes
CIFAR	0.02	Yes
CalTech101	1.87×10^{-6}	Yes
Flowers	0.0032	Yes
Apple Pathology	0.04	Yes

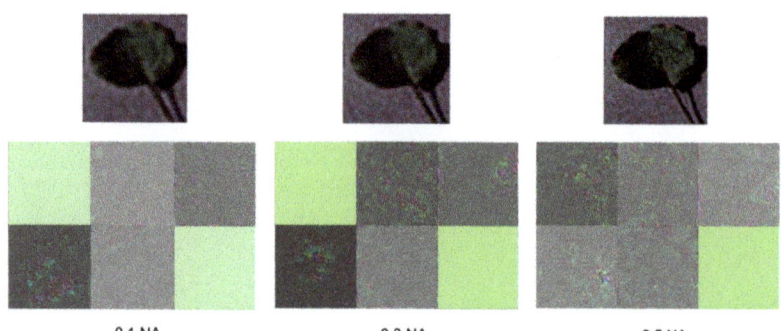

Figure 12. Visualization of the apple pathology images and Google inception convolution neural network activation map (gradient) at the 141st layer. In total, the Google inception convolution neural network had 144 layers. The 141st layer is followed by the fully connected layer, the softmax layer, and the output layer. The green color corresponds to the vanished gradient. Fewer features in the activation map corresponds to the reduction in gradient.

4. Conclusions

In this study, we have investigated image classification accuracy with and without FPM reconstruction with six different image classifiers. We have also compared the image classification accuracy for the FPM reconstructed image versus the lower resolution images shown in Figure 8, Figure 9, Figure 10. It is clear that the lower resolution image has lower image visualization quality than the FPM-reconstructed image. Such a finding is further reinforced by the significant difference between the lower resolution and FPM reconstructed images in terms of image classification accuracy, especially for the lower NA conditions (e.g., 0.05 NA). For MINST, Fashion MNIST, Cifar, Caltech 101, and customized training datasets, when NA is lower, the lower resolution image classification accuracy becomes significantly lower than that of an FPM-reconstructed image. In contrast, when the image quality is higher with a higher NA condition (e.g., 0.5 NA), the image classification performance differs less significantly between with and without FPM reconstruction, which is possibly limited by classifier capabilities.

Nevertheless, the use of FPM reconstruction to improve image classification accuracy is meaningful because the lower NA image without FPM reconstruction suffers from a lower accuracy of image classification. This observation is clear across all six different datasets with different image classifiers. The catastrophic outcome is further supported by the finding that for the Caltech-101dataset, the accuracy of image classification corresponding to lower NA (0.05) and lower resolution conditions is 87% lower than the ground truth. For this situation, the use of FPM reconstruction is indeed helpful, which improves the image classification accuracy from 4.75% to 20%. Our results also suggest that the capabilities of different classifiers differ in terms of their capabilities to deal with lower

NA images. The differences could be caused by different neural network structures. For example, the Google Inception Network is found to have better capabilities to retain the image classification accuracy even for the lower NA images, which is likely caused by the Inception module within the network to handle the lower resolution images.

We further built multiple linear regression between image classification value (dependent variable) and PSNR and SSIM (independent variables). Results show that there is a linear relationship between the dependent variable and independent variables (Figure 8). The related regression F-Test p-value is also smaller than 0.05, which indicates that such a linear relationship is strong. The linear relationship implies that it is feasible to predict the image classification accuracy based on the PSNR and SSIM values. The impact of noises on image classification performance has been documented in previous literature [23,24]. Specifically, Figure 12 shows the activation maps, which are the gradient features learned from the Google inception neural network. It is evident that for the low numerical aperture, which is associated with a lower SSIM and PSNR, more learned gradient features become vanished or reduced. The vanished gradient compromises the network classification accuracy. It is clear that a more vanished gradient in the learned features corresponds to the lower numerical apertures of 0.1 and 0.3, thus compromising network classification accuracy. For a 0.5 numerical aperture, the vanished gradient has been reduced, which, therefore, corresponds to the improved image classification accuracy.

This study had some limitations. First, a greater number of trials for training and the evaluation of image classification accuracy for different NAs are needed. The linear relationship between SSMI, PSNR, and image classification accuracy needs to be evaluated based on a greater number of such trials. Specifically, under extreme lower or higher NA conditions, e.g., lower than 0.05 NA or greater than 0.5 NA, an examination of the relationship is also needed. Furthermore, factors involved in the study are limited. We have not included different noises, training and testing data ratios, or a more extensive number of classifiers. We plan to address the limitations in future work.

Author Contributions: Conceptualization, H.Z., Y.Z., L.W., Z.H., T.-C.P. and P.W.M.T.; methodology, H.Z., W.Z., D.C. and L.W.; software, H.Z. and D.C.; validation, Y.Z., T.-C.P. and P.W.M.T.; formal analysis, H.Z. and T.-C.P.; investigation, T.-C.P.; resources, Y.Z.; data curation, H.Z.; writing—original draft preparation, H.Z. and L.W.; writing—review and editing, T.-C.P., W.Z.; visualization, Y.Z.; supervision, T.-C.P.; project administration, H.Z. and Y.Z.; funding acquisition, Y.Z., P.W.M.T. All authors have read and agreed to the published version of the manuscript.

Funding: National Natural Science Foundation of China (11762009, 61865007); Natural Science Foundation of Yunnan Province (2018FB101); the Key Program of Science and Technology of Yunnan Province (2019FA025); General Research Fund (GRF) of Hong Kong SAR, China (Grant No: 11200319).

Institutional Review Board Statement: Not Applicable.

Informed Consent Statement: Not Applicable.

Data Availability Statement: The data used in this study include MINST, Fashion MNIST, Cifar, Caltech 101. They are available at the following URLs: MINST: http://yann.lecun.com/exdb/mnist/; Fashion MNIST: https://www.kaggle.com/zalando-research/fashionmnist; Cifar: https://www.cs.toronto.edu/~kriz/cifar.html; Caltech 101: http://www.vision.caltech.edu/Image_Datasets/Caltech101/.

Conflicts of Interest: The authors declare no conflict of interest.

References

1. Zheng, G.; Horstmeyer, R.; Yang, C. Wide-field, high-resolution Fourier ptychographic microscopy. *Nat. Photonic* **2013**, *7*, 739–745. [CrossRef] [PubMed]
2. Zheng, G.; Shen, C.; Jiang, S.; Song, P.; Yang, C. Concept, implementations and applications of Fourier ptychography. *Nat. Rev. Phys.* **2021**, *3*, 207–223. [CrossRef]
3. Tian, L.; Li, X.; Ramchandran, K.; Waller, L. Multiplexed coded illumination for Fourier ptychography with an LED array microscope. *Biomed. Opt. Express* **2014**, *5*, 2376–2389. [CrossRef] [PubMed]

4. Sun, J.; Chen, Q.; Zhang, J.; Fan, Y.; Zuo, C. Single-shot quantitative phase microscopy based on color-multiplexed Fourier ptychography. *Opt. Lett.* **2018**, *43*, 3365–3368. [CrossRef] [PubMed]
5. Sun, J.; Zuo, C.; Zhang, J.; Fan, Y.; Chen, Q. High-speed Fourier ptychographic microscopy based on programmable annular illuminations. *Sci. Rep.* **2018**, *8*, 7669. [CrossRef] [PubMed]
6. Jiang, S.; Guo, K.; Liao, J.; Zheng, G. Solving Fourier ptychographic imaging problems via neural network modeling and TensorFlow. *Biomed. Opt. Express* **2018**, *9*, 3306–3319. [CrossRef] [PubMed]
7. Nguyen, T.; Xue, Y.; Li, Y.; Tian, L.; Nehmetallah, G. Deep learning approach for Fourier ptychography microscopy. *Opt. Express* **2018**, *26*, 26470–26484. [CrossRef] [PubMed]
8. Chen, L.; Wu, Y.; Souza, A.M.D.; Abidin, A.Z.; Wismüller, A.; Xu, C. MRI tumor segmentation with densely connected 3D CNN. In Proceedings of the Medical Imaging 2018: Image Processing, (International Society for Optics and Photonics), Houston, TX, USA, 11–13 February 2018; Volume 105741F.
9. Zhu, P.; Wen, L.; Bian, X.; Ling, H.; Hu, Q. Vision meets drones: A challenge. *arXiv* **2018**, arXiv:1804.07437.
10. Goyal, P.; Dollár, P.; Girshick, R.; Noordhuis, P.; Wesolowski, L.; Kyrola, A.; Tulloch, A.; Jia, Y.; He, K. Accurate, Large Minibatch SGD: Training ImageNet in 1 Hour. *arXiv* **2017**, arXiv:1706.02677.
11. Zhang, H.; Zhou, W.-J.; Liu, Y.; Leber, D.; Banerjee, P.; Basunia, M.; Poon, T.-C. Evaluation of finite difference and FFT-based solutions of the transport of intensity equation. *Appl. Opt.* **2017**, *57*, A222–A228. [CrossRef] [PubMed]
12. Zhou, W.-J.; Guan, X.; Liu, F.; Yu, Y.; Zhang, H.; Poon, T.-C.; Banerjee, P.P. Phase retrieval based on transport of intensity and digital holography. *Appl. Opt.* **2018**, *57*, A229–A234. [CrossRef] [PubMed]
13. Zhang, H.; Zhou, W.; Leber, D.; Hu, Z.; Yang, X.; Tsang, P.W.M.; Poon, T.-C. Development of lossy and near-lossless compression methods for wafer surface structure digital holograms. *J. Micro/Nanolithogr. MEMS MOEMS* **2015**, *14*, 41304. [CrossRef]
14. Xiao, H.; Rasul, K.; Vollgraf, R. Fashion-MNIST: A novel image dataset for benchmarking machine learning algorithms. *arXiv* **2017**, arXiv:1708.07747.
15. Zhong, Z.; Zheng, L.; Kang, G.; Li, S.; Yang, Y. Random erasing data augmentation. *arXiv* **2017**, arXiv:1708.04896. [CrossRef]
16. He, K.; Zhang, X.; Ren, S.; Sun, J. Deep residual learning for image recognition. In Proceedings of the IEEE Conference on Computer Vision and Pattern Recognition, Las Vegas, NV, USA, 27–30 June 2016; pp. 770–778.
17. Perez, L.; Wang, J. The effectiveness of data augmentation in image classification using deep learning. *arXiv* **2017**, arXiv:1712.04621.
18. Zhong, Z.; Zheng, L.; Zheng, Z.; Li, S.; Yang, Y. Camera Style Adaptation for Person Re-identification. In Proceedings of the 2018 IEEE/CVF Conference on Computer Vision and Pattern Recognition, Salt Lake City, UT, USA, 18–22 June 2018; pp. 5157–5166.
19. Wang, R.; Song, P.; Jiang, S.; Yan, C.; Zhu, J.; Guo, C.; Bian, Z.; Wang, T.; Zheng, G. Virtual brightfield and fluorescence staining for Fourier ptychography via unsupervised deep learning. *Opt. Lett.* **2020**, *45*, 5405–5408. [CrossRef] [PubMed]
20. Li, Y.; Cheng, S.; Xue, Y.; Tian, L. Displacement-agnostic coherent imaging through scatter with an interpretable deep neural network. *Opt. Express* **2021**, *29*, 2244–2257. [CrossRef] [PubMed]
21. Zhang, H.; Wang, L.; Zhou, W.; Hu, Z.; Tsang, P.; Poon, T.-C. Fourier Ptychography: Effectiveness of image classification. In Proceedings of the SPIE, Melbourne, Australia, 14–17 October 2019; Volume 11205, pp. 112050G-1–112050G-8.
22. Wang, L.; Song, Q.; Zhang, H.; Yuan, C.; Poon, T.-C. Optical scanning Fourier ptychographic microscopy. *Appl. Opt.* **2021**, *60*, A243–A249. [CrossRef] [PubMed]
23. Gowdra, N.; Sinha, R.; MacDonell, S. Examining convolutional feature extraction using Maximum Entropy (ME) and Signal-to-Noise Ratio (SNR) for image classification. In Proceedings of the IECON 2020 the 46th Annual Conference of the IEEE Industrial Electronics Society, Singapore, 18–22 October 2020; pp. 471–476.
24. Li, Q.; Shen, L.; Guo, S.; Lai, Z. Wavelet Integrated CNNs for Noise-Robust Image Classification. In Proceedings of the IEEE/CVF Conference on Computer Vision and Pattern Recognition (CVPR), Seattle, WA, USA, 14–19 June 2020; pp. 7245–7254.

Article

A New Method to Verify the Measurement Speed and Accuracy of Frequency Modulated Interferometers

Toan-Thang Vu [1], Thanh-Tung Vu [1,*], Van-Doanh Tran [1], Thanh-Dong Nguyen [1] and Ngoc-Tam Bui [1,2,*]

[1] School of Mechanical Engineering, Hanoi University of Science and Technology, Hanoi 100000, Vietnam; thang.vutoan@hust.edu.vn (T.-T.V.); doanh.tvcb170290@sis.hust.edu.vn (V.-D.T.); dong.nguyenthanh@hust.edu.vn (T.-D.N.)

[2] Shibaura Institute of Technology, College of Systems Engineering and Science, Tokyo 135-8548, Japan

* Correspondence: tung.vuthanh@hust.edu.vn (T.-T.V.); tambn@shibaura-it.ac.jp (N.-T.B.)

Abstract: The measurement speed and measurement accuracy of a displacement measuring interferometer are key parameters. To verify these parameters, a fast and high-accuracy motion is required. However, the displacement induced by a mechanical actuator generates disadvantageous features, such as slow motion, hysteresis, distortion, and vibration. This paper proposes a new method for a nonmechanical high-speed motion using an electro-optic modulator (EOM). The method is based on the principle that all displacement measuring interferometers measure the phase change to calculate the displacement. This means that the EOM can be used to accurately generate phase change rather than a mechanical actuator. The proposed method is then validated by placing the EOM into an arm of a frequency modulation interferometer. By using two lock-in amplifiers, the phase change in an EOM and, hence, the corresponding virtual displacement could be measured by the interferometer. The measurement showed that the system could achieve a displacement at 20 kHz, a speed of 6.08 mm/s, and a displacement noise level < 100 pm/$\sqrt{\text{Hz}}$ above 2 kHz. The proposed virtual displacement can be applied to determine both the measurement speed and accuracy of displacement measuring interferometers, such as homodyne interferometers, heterodyne interferometers, and frequency modulated interferometers.

Keywords: electro-optic modulator; frequency modulation; displacement measuring interferometer

1. Introduction

High-precision technology is increasingly critical for industrial applications. The demand for high-speed and precise processing has been on the rise in various fields. To meet these requirements, different types of displacement measuring sensors are available, including capacitive sensors, linear encoders, and laser interferometers. In a short measurement range, capacitive sensors can obtain sub-nanometer resolution [1,2]. There are some disadvantages of capacitive sensors, such as sensitivity to temperature and humidity, short stand-off distance, and relatively low bandwidth [3,4]. Linear encoders, which can measure both distance and displacement at nanometer resolution over a long measurement range, are widely used as machine tools [5,6]. However, the complex structure and large volume limits their application in ultra-precision measurements. Among these sensors, displacement measuring interferometers have increasingly been adopted because of their high level of accuracy and traceability to the definition of the meter. Numerous displacement measuring interferometers have been developed, such as homodynes [7–9], heterodynes [10–12], and frequency/phase modulation interferometers [13–16]. In open-air environments, the measurement range and accuracy of the homodyne interferometer are reduced due to the refractive index fluctuation [17]. The heterodyne frequency is less than 20 MHz, and, hence, the measurement speed is approximately 5 m/s [18,19]. For a frequency modulation interferometer, the measurement speed is limited by the modulation frequency of the laser source. Laser diodes (LDs) are widely used as the laser source in interferometers due to their advantageous features, such as high power, compactness, and long lifespan,. In particular, the high modulation frequency of LDs can be obtained by applying a current modulation at the GHz level [20,21]. Hence, the measurement speed of the frequency modulation interferometer can theoretically reach a level of 10 m/s. Therefore, a considerable challenge for the interferometer is the measurement speed verification at nanoscale uncertainty. Normally, a piezo-electric (PZT) actuator uses the motion of the mirrors for interferometry because of its high motion resolution. However, the travelling speed of the PZT stage is limited to several kHz. In addition, a mechanical displacement PZT system typically shows hysteresis, which is a type of cyclic error [22–24]. Hysteresis introduces uncertainties into measurements, which should be suppressed or removed. Voice coil actuators are also used for many high-precision motion applications [25,26]. The actuator can achieve a resolution of 2 nm over a range of 1 mm [27]. Two major disadvantages of voice coil actuators are heat output and disturbance from moving wires [28]. Hence, voice coil actuators are not suitable for wide-range nano-positioning at high speed.

The measurement speed and measurement precision of displacement measuring interferometers are key parameters. For interferometers, the measurement speed or rate implies the time required to take one displacement measurement in the unit of seconds or Hz. In our previous works, a high modulation frequency of 3 MHz was successfully applied to the LD to improve the measurement speed of the frequency modulated interferometer [13,14,29]. Moreover, a high-precision phase meter was developed to measure the phase change in the interferometer [12]. To verify the measurement accuracy of our proposed interferometers, the displacement of the target mirror was measured using a capacitive sensor (D100, Physik Instrumente) integrated into the PZT stage. However, the bandwidth of the capacitive sensor was less than 3kHz, and the resolution was 1 nm over a displacement range of 2 μm [13,14]. Therefore, it remains a significant challenge to verify both parameters using a nanometer resolution and high-speed displacement actuator. In this paper, a high-speed virtual displacement without hysteresis is proposed and validated. It is noteworthy that all types of displacement interferometers determine target displacement by measuring the phase shift. A pure phase modulation at high frequency can be generated using an electro-optic modulator (EOM). This means that the EOM can be used to induce the phase shift rather than mechanical motions, such as PZT or voice coil actuators, to confirm the measurement speed and accuracy of displacement measuring interferometers. By using an EOM, the phase shift can be generated ranging from several kHz to some tens of MHz without distortion, hysteretic, or vibration. Hence, both the measurement accuracy and speed of the interferometer under assessment can be accurately verified. In the primary experiment, an EOM was placed into an arm of a frequency modulated interferometer. The virtual displacement was implemented using the EOM, in which the phase was changed using a modulation voltage at 20 kHz. Using two lock-in amplifiers (LIAs) and a Lissajous diagram, the phase change in the EOM and the corresponding virtual displacement could be measured using the frequency modulated interferometer.

In summary, the main contributions of this paper are listed as follows: (1) the calculation and design of a virtual displacement generated using an EOM; (2) the validation and measurement of the virtual displacement using a frequency modulated interferometer; (3) the comparison of the resulting measurement using the interferometer and theoretical result of the motion. Both the measurement speed and accuracy of the frequency modulated interferometer can be verified using the proposed method. This is significant in high-speed measuring applications, such as spindle error measurement, vibration measurement, and machine tool calibration.

2. Methodology

Figure 1a shows the schematic of the frequency modulated interferometer. The laser source was a laser diode (LD), in which the frequency was modulated with a sinusoidal signal by a modulation current injection. The angle frequency $\omega(t)$ of the LD is expressed as

$$\omega(t) = \omega_0 + \Delta\omega \sin(\omega t) \qquad (1)$$

where ω_0 (=$2\pi f_0$), ω, and $\Delta\omega$ are the initial angle frequency of the LD, acting as the carrier angle frequency, the angle modulation frequency, and the angle modulation frequency excursion of ω, respectively. An isolator was used to prevent unwanted lights returning to the LD. A half-wave plate (HWP) was employed to rotate the polarization plane of the incident light at 45°. A beam splitter (BS) divided the incident light from the isolator into two beams: a reflected beam and a transmitted beam. The reflected beam was sent to a mirror (M1) in the reference arm. In the measurement arm, the second beam was sent to another mirror (M2) along a polarizer (P) and an EOM. The polarizer was placed in front of the EOM to ensure the correct polarization of the input light. After reflection from the mirrors, the reflected lights of the two arms recombined with each other at the BS to generate interference.

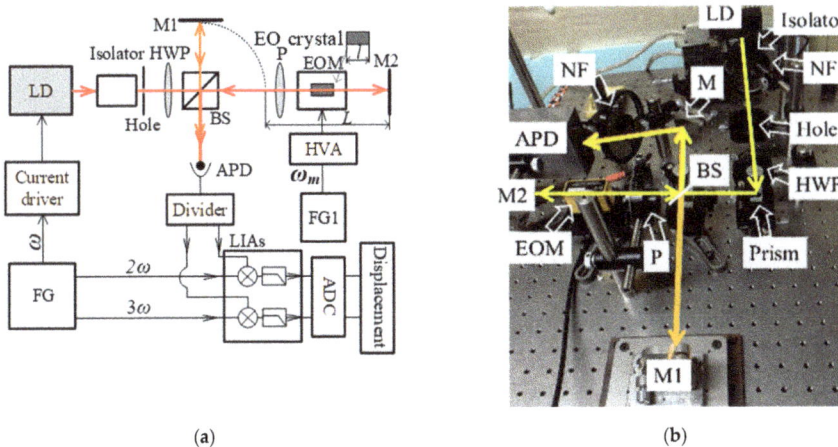

(a) (b)

Figure 1. Virtual displacement-measuring interferometer using a frequency-modulated laser source. (**a**) Schematic design and (**b**) the experimental setup. LD: laser diode; HWP: half-wave plate; BS: beam splitter; M: mirror; P: polarizer; EOM: electro-optic modulator; EO crystal: electro-optic crystal; NF: neutral density filter; HVA: high-voltage amplifier; FG: function generator; APD: avalanche photodetector; LIA: lock-in amplifier; ADC: analog-to-digital converter.

The intensity $I(\tau, t)$ of the interference signal [13,14] is written as

$$I(\tau,t) = E^2_{01} + E^2_{02} + 2E_{01}E_{02}\cos(\omega_0\tau)J_0(m) + 4E_{01}E_{02}\cos(\omega_0\tau)\sum_{n=1}^{\infty}J_{2n}(m)\cos(2n\omega t) \\ -4E_{01}E_{02}\sin(\omega_0\tau)\sum_{n=1}^{\infty}J_{2n-1}(m)\cos[(2n-1)\omega t], \quad (2)$$

where τ, m, E_{01} and E_{02}, n, $J_0(m)$, $J_{2n}(m)$, and $J_{2n-1}(m)$ are the changes in time between the two arms of the interferometer, the modulation index, the amplitudes of the electric fields in the reference and measurement arms, an integer, and the Bessel functions, respectively. Here,

$$m = \frac{\Delta\omega}{\omega}\sin\left(\frac{\omega\tau}{2}\right) \approx \frac{2\pi\Delta f n_{air}L}{c} \quad (3)$$

where Δf ($\Delta\omega = 2\pi\Delta f$), n_{air}, L, and c are the frequency modulation excursion, the refractive index of air, the unbalance length of the interferometer, and the speed of light in a vacuum. A divider split $I(\tau, t)$ into two parts, which are coupled with the two purely sinusoidal signals of 2ω and 3ω from the function generator. By using the two LIAs [13,14], the intensities $I_{2\omega}$ and $I_{3\omega}$ of the 2ω and 3ω harmonics of $I(\tau, t)$, respectively, are produced as follows:

$$I_{2\omega} = 2E_{01}E_{02}\cos(\omega_0\tau)J_2(m), \\ I_{3\omega} = -2E_{01}E_{02}\sin(\omega_0\tau)J_3(m). \quad (4)$$

Here, $J_2(m)$ and $J_3(m)$ are the second- and third-order Bessel functions, respectively. By using the Lissajous diagram, $\omega_0\tau$ is determined. The total phase difference Φ between the arms is

$$\Phi = \omega_0\tau = \arctan\left(-\frac{I_{3\omega}}{I_{2\omega}} \cdot \frac{J_2(m)}{J_3(m)}\right). \tag{5}$$

To generate the virtual displacement, the EOM was applied using a sinusoidal voltage $V(t)$

$$V(t) = V_{EOM}\sin(\omega_{EOM}t), \tag{6}$$

where V_{EOM} is the amplitude of $V(t)$. For the EOM (4002, Newport) [30], the optical phase difference $\Delta\Phi_1$ induced by applying $V(t)$ is

$$\Delta\Phi_1 = \frac{\pi}{V_\pi}V, \tag{7}$$

where V_π is a half-wave voltage. Because the laser beam in the measurement arm was double-phase modulated (Figure 1), by substituting Equation (6) into Equation (7), the phase $\Delta\Phi_{EOM} = 2\Delta\Phi_1$ is

$$\Delta\Phi_{EOM} = 2\Delta\Phi_1 = \frac{2\pi}{V_\pi}V = \frac{2\pi}{V_\pi}V_{EOM}\sin(\omega_{EOM}t). \tag{8}$$

However, the total phase difference Φ between the arms of the interferometer is

$$\Phi = \omega_0\tau = \frac{4\pi n_{air}(L-l)}{\lambda} + \frac{4\pi}{\lambda}(n_e + \Delta n_e)l = \frac{4\pi n_{air}(L-l)}{\lambda} + \frac{4\pi}{\lambda}n_e l + \Delta\Phi_{EOM}, \tag{9}$$

where L, n_{air}, l, n_e, Δn_e, and λ are the unbalance length of the interferometer, the refractive index of air, the length and unperturbed refractive index of the electro-optic (EO) crystal, the change in n_e induced by applying $V(t)$, and the laser wavelength, respectively. Here,

$$\Delta\Phi_{EOM} = \frac{4\pi}{\lambda}\Delta n_e l = \frac{4\pi}{\lambda}\Delta l, \tag{10}$$

Δl is the change in the optical path in the EO crystal, which is defined as the virtual displacement of M2 due to applying $V(t)$. Substituting Equation (8) into Equation (10), Δl becomes

$$\Delta l = \frac{\lambda}{2}\frac{V_0}{V_\pi}\sin(\omega_{EOM}t). \tag{11}$$

Substituting Equations (5) and (10) into Equation (9), Δl measured by the interferometer is

$$\Delta l = \frac{\lambda}{4\pi}\left\{\arctan\left(-\frac{I_{3\omega}}{I_{2\omega}} \cdot \frac{J_2(m)}{J_3(m)}\right)\right\} - [(L-l)n_{air} + n_e l]. \tag{12}$$

In this experiment, we compared the measurement in Equation (12) with the calculated displacement in Equation (11) at the high phase-modulation frequency ω_m.

For the frequency modulated interferometer, the maximum measurable speed V_{max} can be given by [13]

$$V_{max} \leq k\frac{\lambda}{4}f, \tag{13}$$

where k and f are an integer and the modulation frequency of LD, respectively. Here, k represents the ratio of the cutoff frequency of the LIAs and the modulation frequency of LD; normally $k = 0.1$–0.8. In this paper, a virtual displacement was developed at a speed of tens of kHz and without hysteresis generated using an EOM.

3. Experiments and Results

A photograph of the experimental setup is shown in Figure 1b. The laser source was an LD (HL6312G, Thorlabs Inc., Newton, NJ, USA), and it was frequency modulated with a sine-wave signal (a frequency modulation of 20 MHz and modulation excursion of 570 MHz) by modulating the injection current. To produce a high-speed virtual motion, the 4002 EOM with $V_\pi = $ ~125 V at $\lambda = 635$ nm was used. The EOM was phase changed using a 20 kHz sine-wave signal $V(t)$, which had $V_0 = 2$ V, by utilizing a digital function generator (Moku:labs, Liquid Instruments), and it was then amplified 30 times by a high-voltage amplifier. $\Delta\Phi_{EOM}$ was produced, and the 20 kHz sine-wave virtual displacement Δl could thus be attained with a peak-to-peak (p-p) amplitude of ~152 nm (=635 nm × 2 V × 30/2/125 V). The $I(\tau, t)$ was detected using an avalanche photodetector (DET08CFC/M, Thorlabs). To extract $I_{2\omega}$ ($2\omega = 40$ MHz) and $I_{3\omega}$ ($2\omega = 60$ MHz), we used two digital LIAs (Moku:labs, Liquid Instruments), which were synchronized with each other and had a cutoff frequency of 200 kHz. The $I_{2\omega}$ and $I_{3\omega}$ were then recorded using an analog-to-digital converter (AD16-16U(PCI)EV, Contec Co., Osaka, JP) at a sampling frequency of 715 kHz to eliminate MHz-order noises in the harmonics. In this experiment, the modulation index was m ≈ 2.4. This led to $J(2\omega) = 2.15 \times J(3\omega)$. In addition, because $(L-l)n_{air} + n_e l \approx n_{air}L \approx 0.2$ m. Equation (13) is rewritten as

$$\Delta l = \frac{\lambda}{4\pi} \left\{ \arctan\left(-\frac{1}{2.15} \frac{I_{3\omega}}{I_{2\omega}} \right) \right\} - L n_{air}. \tag{14}$$

The experiments were performed under the conditions in Table 1. The second and third harmonics that were clearly detected by LIAs are shown in Figure 2a. A Lissajous diagram using the two harmonics is shown in Figure 2b. A sine-wave displacement at 20 kHz and a p-p amplitude of ~152 nm was obtained over 1 ms, as shown in Figure 3a. This led to a measured speed of 6.08 mm/s ($=2\Delta l \times f_{EOM} = 2 \times 152$ nm $\times 20$ kHz).

Table 1. The experimental conditions for the virtual displacement measurement.

Modulation frequency ω for LD	$2\pi \times 20$ MHz
Frequency modulation excursion $\Delta\omega$	$2\pi \times 570$ MHz
Cutoff frequency of LIAs	200 kHz
V_π at 536 nm	125 V
Applied angle frequency for ω_{EOM}	$2\pi \times 20$ kHz
Unbalanced length L	0.2 m
Working sampling frequency	715 kHz

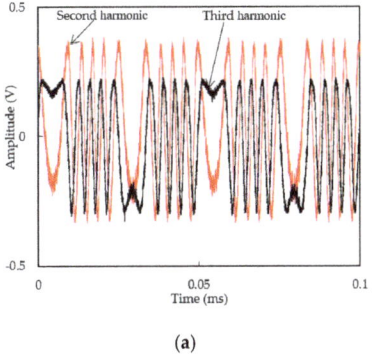 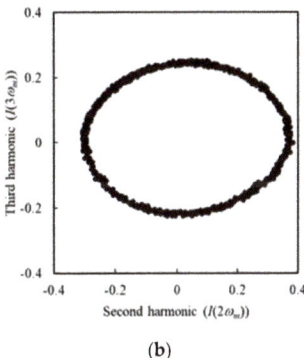

(a) (b)

Figure 2. Harmonic detection: (**a**) second and third harmonics; (**b**) Lissajous diagram using $I_{2\omega}$ and $I_{3\omega}$ harmonics.

The result obtained by the interferometer was compared with a calculated displacement function F using Equation (11), $F = A + B \times \sin(2\pi \times 20 \text{ kHz} \times t)$, where $A = -0.152$ μm and $B = -0.076$ μm. The difference between the measured result and F is shown in Figure 3b. A displacement difference of ~30 nm and a standard deviation of ~6 nm were attained. The displacement noise floor analyzed in Figure 3b is depicted in Figure 3c. Above 2 kHz, a displacement noise floor of less than 100 pm/$\sqrt{\text{Hz}}$ was achieved. A dominated noise peak at 20 kHz can be seen. This led to the cycle error of the sine-wave displacement in Figure 3a, and this noise level was ~100 pm/$\sqrt{\text{Hz}}$. The result showed that our method could generate a stable and high-precision displacement that could be used as a reference. Even though unwanted noises caused by vibration and air disturbance still existed in the measurement, the method was promising for high-speed measurements.

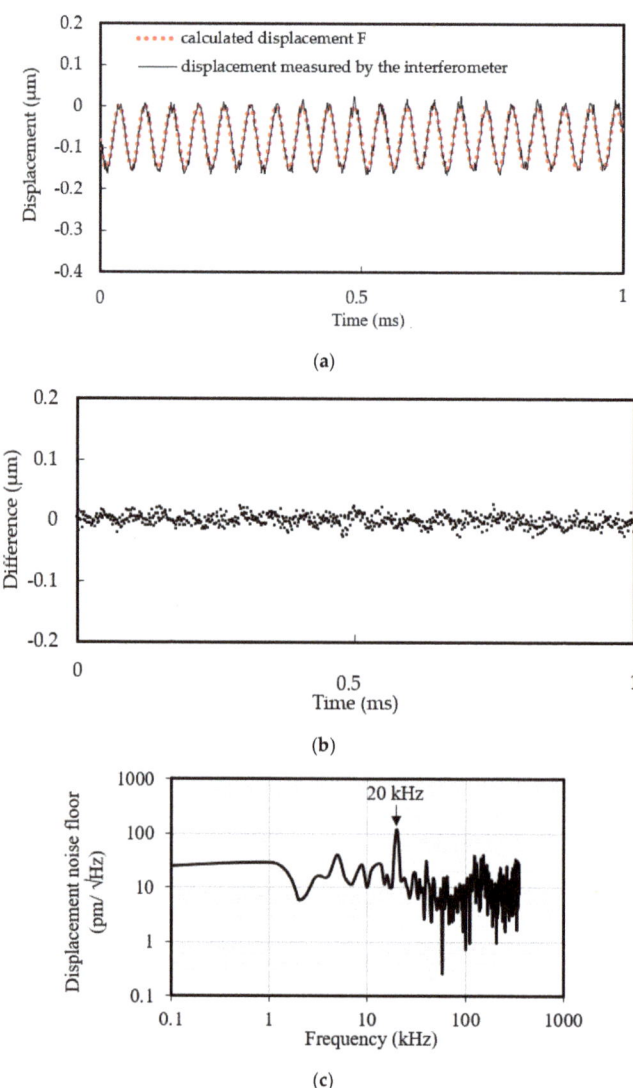

Figure 3. Measurement: (**a**) virtual displacement measured by the interferometer (solid line) compared with the calculated displacement function $F = A + B \times \sin(2\pi \times 20 \text{ kHz} \times t)$ (dot line; see Equation (11), where $A = -0.152$ μm and $B = -0.076$ μm); (**b**) difference between the result obtained by the interferometer and F; and (**c**) displacement noise floor analyzed from (**b**) by Fourier transform.

From Equation (5), when $k = 0.8$, $\lambda = 635$ nm, and $f = 20$ MHz, the maximum measurable speed of 2.54 m/s can be achieved. However, due to the limitation of the hardware caused by the low bandwidth of the high-voltage amplifier (HVA) and LIAs, we could not perform a virtual displacement of several m/s. We plan to use a higher bandwidth HVA and LIAs to improve the measurement speed. Another disadvantage of the proposed method is that the use of the EOM in one arm of the interferometer can cause the polarization mixing effect, and, hence, it may induce some noise of the interference signal. To improve the interference signal, some polarization optics can be used in the next experiment.

4. Conclusions

In this study, high-speed displacement generation using an electro-optic modulator (EOM) was proposed and validated. A frequency modulated interferometer was established to measure the virtual displacement produced by the EOM. To produce a high-speed displacement, an EOM was phase changed using a high-speed sinusoidal modulation voltage. A 20 kHz sine-wave signal with an amplified amplitude of 60 V was applied to the EOM. A corresponding sinusoidal virtual displacement of ~152 nm at 20 kHz was then generated. The interferometer extracted an interference signal that contained the phase change in the EOM. Two LIAs and a Lissajous diagram were adopted for detecting the phase change and calculating the virtual displacement. The measurement showed that the virtual displacement, with an amplitude of 152 nm at 20 kHz and a measurement speed of 6.08 mm/s, was successfully measured by the interferometer. A displacement noise floor < 100 pm/$\sqrt{Hz}$ above 2 kHz was achieved. This experiment was successful at developing a displacement reference at high speed. For future study, the possibility of using a higher modulation for the LDs and the residual amplitude modulation effect will be investigated, as well as high-speed mechanical displacements when the relevant hardware is available.

Author Contributions: Conceptualization, T.-D.N. and T.-T.V. (Thanh-Tung Vu); methodology, T.-T.V. (Toan-Thang Vu); software, V.-D.T.; validation, T.-D.N, and V.-D.T.; formal analysis, N.-T.B.; investigation, T.-T.V. (Thanh-Tung Vu); resources, N.-T.B.; data curation, V.-D.T.; writing—original draft preparation, T.-T.V. (Toan-Thang Vu); writing—review and editing, T.-T.V. (Thanh-Tung Vu); visualization, T.-T.V. (Toan-Thang Vu); funding acquisition, N.-T.B. All authors have read and agreed to the published version of the manuscript.

Funding: This research was funded by Hanoi University of Science and Technology, grant number T2020-PC-201.

Institutional Review Board Statement: Not applicable.

Informed Consent Statement: Not applicable.

Data Availability Statement: This study did not report any data.

Acknowledgments: This work was supported by the Centennial SIT Action for the 100th anniversary of Shibaura Institute of Technology to enter the top ten Asian Institute of Technology.

Conflicts of Interest: The authors declare no conflict of interest.

References

1. Peng, K.; Yu, Z.; Liu, X.; Chen, Z.; Pu, H. Features of capacitive displacement sensing that provide high-accuracy measurements with reduced manufacturing precision. *IEEE Trans. Ind. Electron.* **2017**, *64*, 7377–7386. [CrossRef]
2. Ye, Y.; Zhang, C.; He, C.; Wang, X.; Huang, J.; Deng, J. A review on applications of capacitive displacement sensing for capacitive proximity sensor. *IEEE Access* **2020**, *8*, 45325–45342. [CrossRef]
3. Chi, C.; Sun, X.; Xue, N.; Li, T.; Liu, C. Recent progress in technologies for tactile sensors. *Sensors* **2018**, *18*, 948. [CrossRef]
4. Zhang, D.; Zhao, S.; Zheng, Q.; Lin, L. Absolute capacitive grating displacement measuring system with both high-precision and long-range. *Sens. Actuators A Phys.* **2019**, *295*, 11–22. [CrossRef]
5. Ye, G.; Liu, H.; Ban, Y.; Shi, Y.; Yin, L.; Lu, B. Development of a reflective optical encoder with submicron accuracy. *Opt. Commun.* **2018**, *411*, 126–132. [CrossRef]
6. Khouygani, M.H.G.; Jeng, J.Y. High-precision miniaturized low-cost reflective grating laser encoder with nanometric accuracy. *Appl. Opt.* **2020**, *59*, 5764–5771. [CrossRef]
7. Hori, Y.; Gonda, S.; Bitou, Y.; Watanabe, A.; Nakamura, K. Periodic error evaluation system for linear encoders using a homodyne laser interferometer with 10 picometer uncertainty. *Precis. Eng.* **2018**, *51*, 388–392. [CrossRef]
8. Yan, L.; Chen, B.; Chen, Z.; Xie, J.; Zhang, E.; Zhang, S. Phase-modulated dual-homodyne interferometer without periodic nonlinearity. *Meas. Sci. Technol.* **2017**, *28*, 115006. [CrossRef]
9. Lou, Y.; Yan, L.; Chen, B. A phase modulating homodyne interferometer with tilting error compensation by use of an integrated four-photodetector. *Rev. Sci. Instrum.* **2019**, *90*, 025111. [CrossRef]
10. Joo, K.N.; Clark, E.; Zhang, Y.; Ellis, J.D.; Guzmán, F. A compact high-precision periodic-error-free heterodyne interferometer. *JOSA A* **2020**, *37*, B11–B18. [CrossRef]
11. Yokoyama, S.; Hori, Y.; Yokoyama, T.; Hirai, A. A heterodyne interferometer constructed in an integrated optics and its metrological evaluation of a picometre-order periodic error. *Precis. Eng.* **2018**, *54*, 206–211. [CrossRef]

12. Nguyen, T.D.; Duong, Q.A.; Higuchi, M.; Vu, T.T.; Wei, D.; Aketagawa, M. 19-picometer mechanical step displacement measurement using heterodyne interferometer with phase-locked loop and piezoelectric driving flexure-stage. *Sens. Actuators A Phys.* **2020**, *304*, 111880. [CrossRef]
13. Vu, T.T.; Higuchi, M.; Aketagawa, M. Accurate displacement-measuring interferometer with wide range using an I2 frequency-stabilized laser diode based on sinusoidal frequency modulation. *Meas. Sci. Technol.* **2016**, *27*, 105201. [CrossRef]
14. Duong, Q.A.; Vu, T.T.; Higuchi, M.; Wei, D.; Aketagawa, M. Iodine-frequency-stabilized laser diode and displacement-measuring interferometer based on sinusoidal phase modulation. *Meas. Sci. Technol.* **2018**, *29*, 065204. [CrossRef]
15. Zhang, S.; Yan, L.; Chen, B.; Xu, Z.; Xie, J. Real-time phase delay compensation of PGC demodulation in sinusoidal phase-modulation interferometer for nanometer displacement measurement. *Opt. Express* **2017**, *25*, 472–485. [CrossRef]
16. Xu, J.; Huang, L.; Yin, S.; Gao, B.; Chen, P. All-fiber self-mixing interferometer for displacement measurement based on the quadrature demodulation technique. *Opt. Rev.* **2018**, *25*, 40–45. [CrossRef]
17. Jang, Y.S.; Kim, S.W. Compensation of the refractive index of air in laser interferometer for distance measurement, A review. *Int. J. Precis. Eng. Manuf.* **2017**, *18*, 1881–1890. [CrossRef]
18. Demarest, F.C. High-resolution, high-speed, low data age uncertainty, heterodyne displacement measuring interferometer electronics. *Meas. Sci. Technol.* **1998**, *9*, 1024. [CrossRef]
19. Topcu, S.; Chassagne, L.; Haddad, D.; Alayli, Y.; Juncar, P. Heterodyne interferometric technique for displacement control at the nanometric scale. *Rev. Sci. Instrum.* **2003**, *74*, 4876–4880. [CrossRef]
20. Chan, S.C.; Liu, J.M. Frequency modulation on single sideband using controlled dynamics of an optically injected semiconductor laser. *IEEE J. Quantum Electron.* **2006**, *42*, 699–705. [CrossRef]
21. D'Amato, F.; De Rosa, M. Tunable diode lasers and two-tone frequency modulation spectroscopy applied to atmospheric gas analysis. *Opt. Lasers Eng.* **2002**, *37*, 533–551. [CrossRef]
22. Yan, P.; Zhang, Y. High precision tracking of a piezoelectric nano-manipulator with parameterized hysteresis compensation. *Smart Mater. Struct.* **2018**, *27*, 065018. [CrossRef]
23. Cai, K.; Tian, Y.; Wang, F.; Zhang, D.; Liu, X.; Shirinzadeh, B. Modeling and tracking control of a novel XYθz stage. *Microsyst. Technol.* **2017**, *23*, 3575–3588. [CrossRef]
24. Lin, C.; Yu, J.; Wu, Z.; Shen, Z. Decoupling and control of micromotion stage based on hysteresis of piezoelectric actuation. *Microsyst. Technol.* **2019**, *25*, 3299–3309. [CrossRef]
25. Teo, T.J.; Chen, I.M.; Yang, G.; Lin, W. A flexure-based electromagnetic linear actuator. *Nanotechnology* **2008**, *19*, 315501. [CrossRef]
26. Youm, W.; Jung, J.; Lee, S.; Park, K. Control of voice coil motor nanoscanners for an atomic force microscopy system using a loop shaping technique. *Rev. Sci. Instrum.* **2008**, *79*, 13706–13707. [CrossRef]
27. Fukada, S.; Nishimura, K. Nanometric positioning over a one-millimeter stroke using a flexure guide and electromagnetic linear motor. *Int. J. Precis. Eng. Manuf.* **2007**, *8*, 49–53.
28. Kordonskii, V.; Demchuk, S. Heat transfer in electrodynamic transducers. *J. Eng. Phys.* **1990**, *59*, 1499–1504. [CrossRef]
29. Vu, T.T.; Maeda, Y.; Aketagawa, M. Sinusoidal frequency modulation on laser diode for frequency stabilization and displacement measurement. *Measurement* **2016**, *94*, 927–933. [CrossRef]
30. Newport. DC-250 MHz Electro-Optic Phase Modulator–Models 400X. U.S. Patent 5,189,547, 26 April 2021.

Communication

Integrated IR Modulator with a Quantum Cascade Laser

Janusz Mikołajczyk * and Dariusz Szabra

Institute of Optoelectronics, Military University of Technology, S.Kaliskiego 2, 00908 Warsaw, Poland; dariusz.szabra@wat.edu.pl
* Correspondence: janusz.mikolajczyk@wat.edu.pl; Tel.: +48-261-839-792

Abstract: This paper presents an infrared pulsed modulator into which quantum cascade lasers and a current driver are integrated. The main goal of this study was to determine the capabilities of a new modulator design based on the results of its electrical model simulation and laboratory experiments. A simulation model is a unique tool because it includes the electrical performance of the lasing structure, signal wiring, and driving unit. In the laboratory model, a lasing structure was mounted on the interfacing poles as close to the switching electronics as possible with direct wire bonding. The radiation pulses and laser biasing voltage were registered to analyze the influence of laser module impedance. Both simulation and experimental results demonstrated that the quantum cascade laser (QC laser) design strongly influenced the shape of light, driving current, and biasing voltage pulses. It is a complex phenomenon depending on the laser construction and many other factors, e.g., the amplitude and time parameters of the supplying current pulses. However, this work presents important data to develop or modify numerical models describing QC laser operation. The integrated modulator provided pulses with a 20–100 ns duration and a frequency of 1 MHz without any active cooling. The designed modulator ensured the construction of a sensor based on direct laser absorption spectroscopy, applying the QC laser with spectral characteristics matched to absorption lines of the detected substances. It can also be used in optical ranging and recognition systems.

Keywords: quantum cascade laser; laser controller; infrared modulator; laser spectroscopy; free space optics

1. Introduction

Quantum cascade (QC) laser technology is a commonly used radiation source in many applications that operates in mid-and far-infrared wavelengths [1–3]. These lasers are mainly used in chemical sensing and spectroscopy, in which low noise and high stability of both the light power and spectra are required [4,5]. To this application, a special low-noise current driver and temperature stabilizing unit were constructed [6]. However, these lasers are also being used in other applications as a result of their features, e.g., high power, modulation bandwidth, direct current control, compact size, room-temperature operation, and operation in high transparent spectral ranges of the atmosphere. These features are important for telecommunication, security, and defense technologies [7,8].

Some research works of the modeling techniques used for a simulation of QC lasers have been conducted. Their main focus was to describe the physical processes in electron transport in multiple-quantum-well heterostructures to improve properties for operating temperature, efficiency, and spectral range [9].

What is more, some electrical equivalent circuits were also developed to simulate the laser operation, analyzing the laser electrical features and driving electronics. These works use an emulator of simulating electrical circuits without creating additional mathematical functions or numerical calculations [10]. It is not so computationally intensive and can be implemented in lasing structure optimization or an applied-level device.

Although a simulation model of a realistic circuit should consider five rate equations of quantum wells, three-level or two-level models with certain approximations have been used [10–12]. Based on these models, the light-current characteristics of QC laser were analyzed. These simulations were limited to intensity modulation characteristics for a different ratio to the threshold current.

However, only a few papers describe a QC laser pulse response [13]. The obtained results were used to define the influence of different bias currents on the turn-on delay of the pulse current conversion to the light signal. During described simulations, no model of the current driving unit was applied. The results verification was based on data from other circuit models or numerical calculations. These can be critical for some aspects of applied science in which laser-based systems are analyzed.

QC lasers require a new class of pulse drivers because their operational characteristics are different from those of a laser diode [14]. These lasers need driving signals (current and voltage) of an order higher than those of a bipolar diode laser [15]. The high voltage results from the applied cascade construction, in which the energy of the emitted photons is multiplied by the number of stages (e.g., 30–50). It can reach 20 V or more. The applied intersubband transitions are characterized by a lower optical gain than the interband, requiring high supplying currents. Therefore, the threshold current density of QC lasers is more than one order higher than that of diode lasers [16]. These features represent a challenge when constructing a laser driver that can consider both current and voltage. In these devices, various methods are applied to connect the signals to the QC laser. The most popular is to design unique signal poles or sockets on the printed circuit board (PCB) to connect the pins of the laser housing, e.g., butterfly, transistor outline package (TO), high heat load housing (HHL), and others. However, in a short pulse transmission, the main issues are impedance matching and inductance minimization. However, the matching can be impossible, because laser impedance depends on its construction and fabrication, temperature, or level of driving current [17]. The most critical is its dynamic change during current switching.

The properties of QC laser modulator radiation are matched to its application. Most often, they concern a required shape and the power of the emitted light. In the case of gas sensing devices, the applied laser absorption spectroscopy (LAS) technique determines the optical signal shape. This shape has a form ranging from a simple pulse to complex waveforms. For example, a QC laser pulse with a time duration of 500 ns was used in an intra-pulsed tuned LAS to detect the spectra of nitrogen oxide (NO) and nitrous oxide (N_2O) at around 5.25486 µm and 4.52284 µm, respectively [18]. Shorter pulses are also used, for example, in cavity-enhanced absorption spectroscopy to detect ammonia (NH_3), with a QC laser emitting 50 ns pulses at 200 kHz frequency around 6.8 µm [19].

More complex optical signals are preferable in a tunable laser absorption spectroscopy (TLAS). To obtain scanning of an absorption line range, the QC laser spectra are tuned applying both current ramp signal and high-frequency wave. Many LAS setups use tunable QC lasers, e.g., direct absorption spectroscopy or wavelength modulation [20]. In these setups, IR modulators generated the current signal consisting of, for example, a slow ramp (10 Hz) and a sinusoid (8 kHz) [21].

A high rate of radiation modulation is the primary goal of IR modulator construction for free-space optics (FSO). Some experimental results with GHz- range modulation of QC lasers have been described. In those tests, some specific architectures of laser structure were used [22]. For example, a laser waveguide with a microstrip line ensured a modulation rate of up to 14 GHz [23]. However, a 'high speed' technology of QC lasers is underdeveloped. The modulation speeds up to 26.5 GHz were obtained using a special design QC laser placed on a cold finger with continuous-flow liquid nitrogen and driven with bias-tee and a microwave signal [24].

There are many other applications for which it is important to design a compact and low-power consumption IR modulator. For example, it could be used as an alarm beacon with optical signals detected at extended ranges, a jamming device to disorientate an incoming missile threat, or a light source for beam riding and ranging systems. There are only a few such devices described in the literature [25]. However, they can work in pulsed mode with closed construction, allowing light pulse generation in strictly defined pulse configurations at the maximum repetition rate of 500 kHz.

This study describes an integrated IR pulsed modulator consisting of a QC laser structure and a current switching module. It ensured the generation of short light pulses (tens of ns) with a maximum frequency of 1 MHz, using direct wire bonding of the lasing structure and the current terminals. Various simulations of this configuration were performed considering different QC laser electrical parameters. Finally, the preliminary experiments were performed with the designed modulator. The obtained results provided experimental verification of, e.g., the influence of both QC laser parameters and the driving signal interface on the light pulses, determination of the electrical scheme of the QC laser, and the integrated modulator. In the future, this modulator can be installed in some standard housings, e.g., HHL, butterfly.

2. QC Compact Laser Modulator Design

A compact laser modulator was built using a laser switching high-speed iC-G30 iCSY HG20M module (iC-House Corporation, Bodenheim, Germany) and a laser structure designed at the Institute of Electron Technology (Warsaw, Poland) [26,27]. The module is equipped with six channels, providing a speed of up to 250 MHz, current pulses of up to 5 A (per channel), and an output voltage of 30 V. Its main capabilities are independent and simple voltage control of channel currents, parallel channel operation of up to 6 A (constant mode) and up to 30 A (pulsed mode), and thermal shutdown. The QC laser was designed to operate at room temperature. Its parameters and construction mean it can be mounted directly on the signal pole of the laser switching module without the need to use an extra cooling unit. The optical power and voltage vs. current characteristics of the laser and a picture of the integrated modulator are presented in Figure 1.

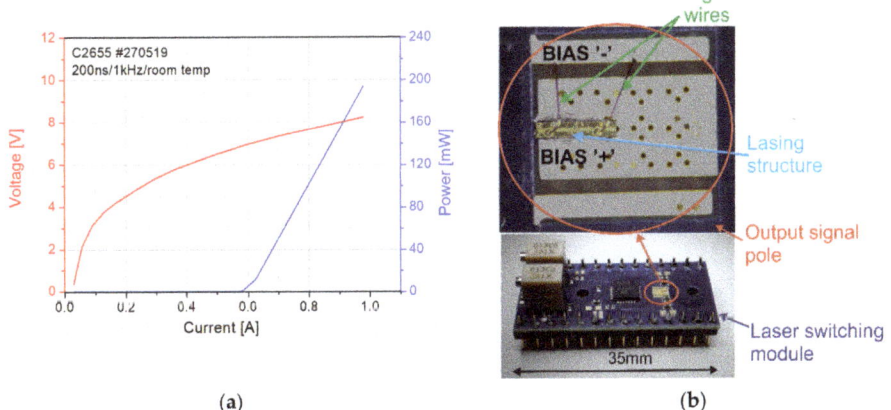

Figure 1. Optical power and voltage vs. current of the mounted QC laser (**a**) and a picture of the integrated IR modulator (**b**).

3. Simulations

The pulse operation performances of the laser modulator were analyzed using SPICE (Simulation Program with Integrated Circuit Emulation) models of the switching module and the prepared electrical model of the QC laser (LTSpice XVII software ver.17.0.27.0, Analog Devices, Norwood, MA, USA). During the simulations, no parameters for the PCB technology were included. Such assumptions simplify the scope of simulations but should be considered in actual conditions [28]. Its manufacturer delivered the SPICE description of the switching module. For the QC laser, the SPICE model was prepared using data from Alpes Lasers (St Blaise, Switzerland) for the lasing structure, and IXYS Corporation (Milpitas, CA, USA) for the inductance of the laser package [29,30]. In Figure 2, an electrical schematic of the IR lighting device is presented. The Alpes laser model combines a resistor (R_L) and two capacitors (C_1, C_2). The R_L values depend on the laser operation point (e.g., for a low current: 10–20 Ω, and for the lasing current: 1–3 Ω). The C_1 (~100 pF) and C_2 (below 100 pF) are mainly determined by bonding pads and laser mounting technologies, respectively. IXYS Corporation determines the inductance level (L) for various signal interface technologies applied in laser housings. For example, a connection of two points at a distance of 0.2 mm, with a copper wire 0.014 mm in diameter and a height of 0.06 mm, gives an inductance of 3.6 nH.

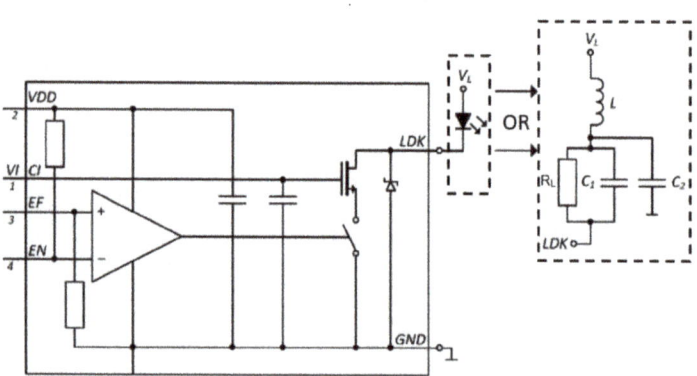

Figure 2. Electrical scheme of the IR modulator (V_{DD}—voltage supply of the module, VI (CI)—voltage control of output current, EF-EN—differential triggering signals).

A simulation of the light device was performed for different lasing structure electrical parameters (Figure 3). The reactance parts of load impedance did not influence the pulse amplitude (Figure 3a,b). The capacitances generate some oscillations at the pulse rising edge caused by feedback signals conditions. Growing these capacitances, the response time constant also increases. The most critical issue is load inductance, considering pulse shape and its oscillations. We observed a slower modulator response and current high-level changes for both biasing ranges. A current-voltage dependence for the inductor and signal resonance conditions defined these effects. Even at the rising and falling parts of the pulse, these signal fluctuations can be the critical point for the current amplitude, causing, e.g., laser structure damage. The bonding pad capacitance is an almost ideal plate capacitor with a PTFE dielectric [31]. For inductance, the bonding wire generally is 1 nH/mm. However, all of these parameters vary from laser to laser. To reduce pulse degradation, i.e., overshoot and ringing, any unnecessary hardware between the switching unit and the laser must be avoided. That is why the laser's direct and short connection to the electronics is preferable [32].

The load resistance has a direct influence on current amplitude and response time constant (Figure 3c). High resistance attenuates signal fluctuations at the pulse rising edge and increases its fall time. The current level is inversely proportional to this resistance. This is a crucial issue with regard to QCL's switching with dynamic resistance. There is a need to supply lasing structure to ensure population inversion (high resistance—low current—no light) and to emit photons (low resistance—high current—light) [33].

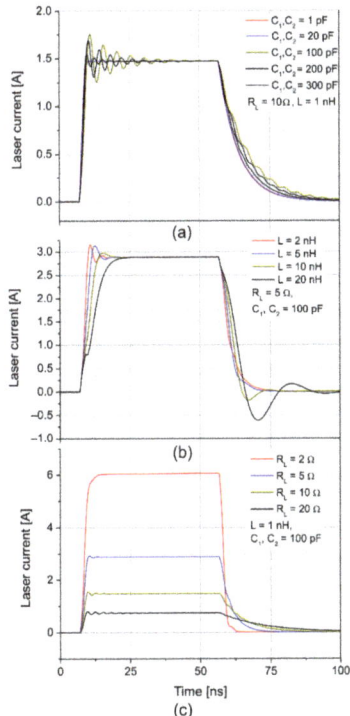

Figure 3. Simulation results of the modulator for different lasing structure electrical parameters: capacitance (**a**), inductance (**b**), and resistance (**c**).

4. Tests of the Laser Switching Module

The switching module tests were performed to analyze the influence of the load resistance on the shapes of the generated current pulses from the switching module using a special test board. The electrical signals were registered using the same scope Tektronix MSO 6 with the current probe CT-1 and the active differential voltage TDP-1500 probe. Figure 4 presents both the current and voltage signals for the two load resistances. These resistances were measured using a Keithley 236 source-measure unit. The module was operating in the current source region, where the current depends only on the 'load' of the biasing voltage. In this region, its output resistance is lower than 300 mΩ.

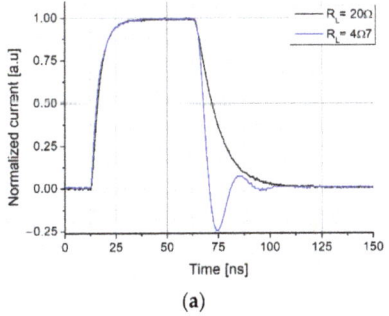
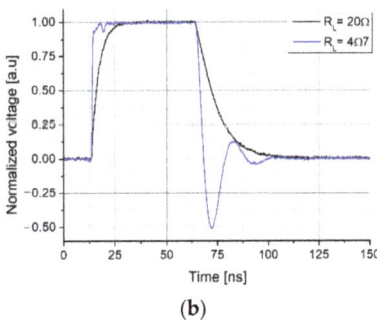

Figure 4. Normalized electrical driving signals: current (**a**) and voltage (**b**).

We observed both voltage and current oscillations at the falling slope for low load resistance. Impedance mismatching caused the transmission and reflection effects in the signal line. An increase in load resistance allowed for the minimizing of these effects with the limited bandwidth.

The load resistance influenced both edges of the voltage signals, but for the current, that was only noticed for the falling one. These differences can modulate the lasing conditions of the QC structure as determined by both current and voltage signals. In practice, these conditions are difficult to predict because they are related to the change in the QC laser impedance during its operation. Therefore, both the signal interface and QC laser parameters form the shapes of the light pulses.

5. Laboratory Tests of the IR Modulator

The preliminary tests of the IR modulator were performed using the lab setup presented in Figure 5. The modulator was placed on the test board to supply all signals (the voltage supply of the switching module, the differential triggering signals, and the biasing of the QC laser). The device was switched using the AFG 3252 model generator (Tektronix, Beaverton, OR, USA). The light pulses were registered with the PVI-4TE-10.6 (VIGO System S.A., Warsaw, Poland) detection module with a responsivity of 2.5×10^4 V/W and bandwidth of 500 MHz. An oscilloscope of the MSO 6 model (Tektronix) with the active probe (TDP-1500, Tektronix) visualized the voltage biasing of the laser structure. There was no technical possibility of registering current pulses (the laser wire was directly bonded on the driver signal pole) and analyzing their shapes.

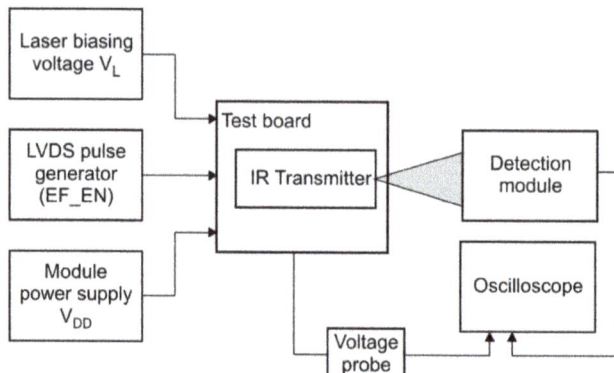

Figure 5. View of the testing setup and lasing device placed on the test board irradiating detection module.

Figure 6 presents some shapes of the registered voltage signals and radiation pulses for different pulse time durations at a frequency of 1 MHz. The voltage signal came ~5 ns faster than the light one, and the laser biasing voltage limited its amplitude. It agrees strongly with the data described in [28].

Some fluctuations caused by impedance mismatching of the signal interface were also observed. The high similarity of the shape of voltage signals obtained during reference measurements using a resistor (Figure 4b) and the QC laser is notable. But these shapes differed from those of the light pulses. A slower rising edge and laser pulse oscillations were noticed.

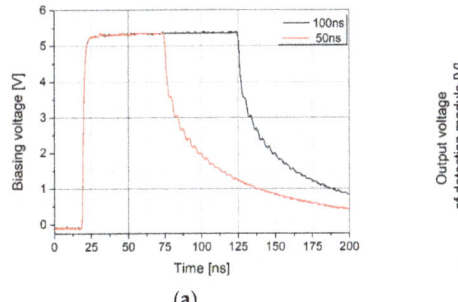
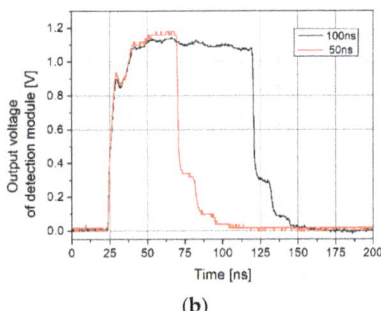

Figure 6. Registered signals for different pulse durations: biasing voltage (**a**) and radiation pulses (**b**).

Some virtues of the experimental and simulation results are shared, e.g., inflection in the rising slope, oscillations at the top signal, and some peaks at the falling edge (Figure 7). These effects were not registered for the tests with resistances, indicating that their source is a powered laser structure. This is a new aspect for analyses of QC laser module construction for applied science. Experimental results of laser biasing voltage and light pulse also defined the time delay of ~5 ns. This phenomenon was analyzed during modulation simulations.

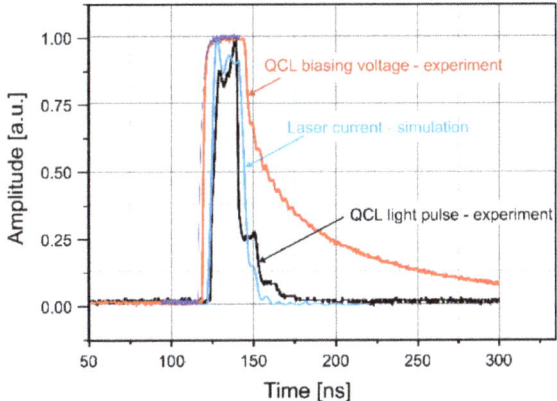

Figure 7. Comparison of experimental results (laser biasing voltage and light pulse) with simulation results (laser current).

6. Conclusions

This study presents the preliminary test of an integrated IR modulator. A unique characteristic of this study was the simulation of the electrical signals supplying the quantum cascade lasers, and the integration of the lasing structure and the current switching module. For this purpose, an electric circuit for the lasing structure was proposed. The simulated results defined the influence of impedance mismatching on both the current and voltage supply signals. It was shown that both signal interfaces and laser parameters form the shape of light pulses. The QCL impedance has non-Ohmic character. That is why it requires more advanced models to describe the behavior of the laser. A perfect impedance match is also impossible, because the QCL impedance changes with the modulation of the driving signal.

Finally, the integrated IR modulator was constructed. The shapes of its light pulses were comparable to the driving current obtained during simulations. These results confirm the phenomenon of dynamic changes in QC laser impedance with current pulse duration. The electrical circuit emulation does not ensure the observation of these results. However, they provide new knowledge in the field of modeling QC lasing structures considering the laser-system approach.

The practical result of the work is an IR modulator that generates MWIR pulses using a QC laser operated at room temperature. The light time parameters, time duration of few tens of ns and max. frequency of 1 MHz, are unique results considering the performance of the other available compact QC laser modules.

Theoretically, the applied switching module also provides a bandwidth of 250 MHz and the generation of complex waveforms using six independently controlled current signals of up to 5 A, and a biasing voltage of 30 V. In the future, it will give new opportunities for many advanced applications. Such a modulator with combined optical signals is needed, e.g., in tunable direct laser absorption spectroscopy, multi-level optical signal transmission, and ultra-short pulse generation with a pre-biasing current.

Author Contributions: Conceptualization, J.M.; methodology, J.M. and D.S.; validation, J.M.; formal analysis, J.M.; investigation, D.S.; resources, J.M.; data curation, J.M.; writing—original draft preparation, J.M.; writing—review and editing, J.M. and D.S.; visualization, D.S. Both authors have read and agreed to the published version of the manuscript.

Funding: Research funded by Narodowe Centrum Badań i Rozwoju Grant No MAZOWSZE/0196/19-00) and by Military University of Technology (Grant no UGB/22-786/2020/WAT).

Institutional Review Board Statement: Not applicable.

Informed Consent Statement: Not applicable.

Data Availability Statement: Not applicable.

Acknowledgments: I would like to thank for technical support with Zbigniew Zawadzki from the Institute of Optoelectronics, MUT. Assistance of the Sieć Badawcza Łukasiewicz—Instytut Technologii Elektronowej (Kamil Pierściński) in providing support of quantum cascade lasers technology is gratefully acknowledged.

Conflicts of Interest: The authors declare no conflict of interest. The funders had no role in the design of the study; in the collection, analyses, or interpretation of data; in the writing of the manuscript, or in the decision to publish the results.

References

1. Pecharroman-Gallego, R. An overview on quantum cascade lasers: Origins and development. In *Quantum Cascade Lasers*; Stavrou, V.N., Ed.; InTech: London, UK, 2017; pp. 1–24. [CrossRef]
2. Patel, C.K.N.; Lyakh, A.; Maulini, R.; Tsekoun, A.; Tadjikov, B. QCL as a game changer in MWIR and LWIR military and homeland security applications. *SPIE Proc.* **2012**, *8373*, 1–9. [CrossRef]
3. Bielecki, Z.; Stacewicz, T.; Wojtas, J.; Mikołajczyk, J. Application of quantum cascade lasers to trace gas detection. *Bull. Polish Acad. Sci. Tech. Sci.* **2015**, *63*, 515–525. [CrossRef]
4. Schwaighofer, A.; Brandstetter, M.; Lendl, B. Quantum cascade lasers (QCLs) in biomedical spectroscopy. *Chem. Soc. Rev.* **2017**, *46*, 5903–5924. [CrossRef] [PubMed]
5. Tombez, L.; Schilt, S.; Di Francesco, J.; Führer, T.; Rein, B.; Walther, T.; Di Domenico, G.; Hofstetter, D.; Thomann, P. Linewidth of a quantum-cascade laser assessed from its frequency noise spectrum and impact of the current driver. *Appl. Phys. B* **2012**, *109*, 407–414. [CrossRef]
6. Taubman, M.S. Low-noise high-performance current controllers for quantum cascade lasers. *Rev. Sci. Instrum.* **2011**, *82*, 1–8. [CrossRef]
7. Grasso, R.J. Defence and security applications of quantum cascade lasers. *SPIE Proc.* **2016**, *99330*, 1–12. [CrossRef]
8. Mikołajczyk, J. An overview of free space optics with quantum cascade lasers. *Int. J. Electron. Telecommun.* **2014**, *60*, 259–264. [CrossRef]
9. Jirauschek, C.; Kubis, T. Modeling techniques for quantum cascade lasers. *Appl. Phys. Rev.* **2014**, *1*. [CrossRef]
10. Yong, K.; Haldar, M.; Webb, J.F. An equivalent circuit for quantum cascade lasers. *J. Infrared Milli Terahz Waves* **2013**, *34*. [CrossRef]
11. Donovan, K.; Harrison, P.; Kelsall, R.W. Self-consistent solutions to the intersubband rate equations in quantum cascade lasers: Analysis of a GaAs/AlxGa1−xAs device. *J. Appl. Phys.* **2001**, *89*, 3084. [CrossRef]
12. Darman, M.; Fasihi, K. Circuit-level modeling of quantum cascade lasers: Influence of Kerr effect on static and dynamic responses. *Optik* **2016**, *127*, 10303–10310. [CrossRef]

13. Chen, G.C.; Fan, G.H.; Li, S.T. Spice simulation of a large-signal model for quantum cascade laser. *Opt. Quant Electron.* **2008**, *40*, 645–653. [CrossRef]
14. Lindquist, J.R. Laser Drivers: Using a Laser Diode or Quantum-Cascade Laser? Don't Forget the Electronics. Laser Focus World 2018. Available online: https://digital.laserfocusworld.com/laserfocusworld/201806/MobilePagedArticle.action?articleId=1404129#articleId1404129 (accessed on 12 March 2021).
15. Zhang, Y.G.; Gu, Y.; Li, Y.Y.; Li, A.Z.; Li, C.; Cao, Y.Y.; Zhou, L. An effective TDLS setup using homemade driving modules for evaluation of pulsed QCL. *Appl. Phys. B* **2012**, *109*, 541–548. [CrossRef]
16. Tournié, E.; Baranov, A.N. Mid-infrared semiconductor lasers: A review. In *Semiconductors and Semimetals*; Coleman, J.J., Bryce, A.C., Jagadishm, C., Eds.; Elsevier: Amsterdam, The Netherlands, 2012; Volume 86, pp. 183–226. [CrossRef]
17. Ashok, P.; Ganesh, M. Impedance characteristics of mid infra red quantum cascade lasers. *Opt. Laser Technol.* **2021**, *134*. [CrossRef]
18. Douat, C.; Hübner, S.; Engeln, R.; Benedikt, J. Production of nitric/nitrous oxide by an atmospheric pressure plasma jet. *Plasma Sources Sci. Technol.* **2016**, *25*. [CrossRef]
19. Gadedjisso-Tossou, K.S.; Stoychev, L.I.; Mohou, M.A.; Cabrera, H.; Niemela, J.; Danailov, M.B.; Vacchi, A. Cavity ring-down spectroscopy for molecular trace gas detection using a pulsed DFB QCL emitting at 6.8 µm. *Photonics* **2020**, *7*, 74. [CrossRef]
20. Bolshov, M.A.; Kuritsyn Yu, A.; Romanovskii Yu, V. Tunable diode laser spectroscopy as a technique for combustion diagnostics. *Spectrochim. Acta Part B At. Spectrosc.* **2015**, *106*, 45–66. [CrossRef]
21. Upadhyay, A.; Wilson, D.; Lengden, M.; Chakraborty, A.L.; Stewart, G.; Johnstone, W. Calibration-free WMS using a cw-DFB-QCL, a VCSEL, and an Edge-emitting DFB laser with in-situ real-time laser parameter characterization. *IEEE Photonics J.* **2017**, *9*. [CrossRef]
22. Pirotta, S.; Tran, N.-L.; Jollivet, A.; Biasiol, G.; Crozat, P.; Manceau, J.-M.; Bousseksou, A.; Colombelli, R. Fast amplitude modulation up to 1.5 GHz of mid-IR free-space beams at room-temperature. *Nat. Commun.* **2021**, *12*. [CrossRef]
23. St-Jean, M.R.; Amanti, M.I.; Bernard, A.; Calvar, A.; Bismuto, A.; Gini, E.; Beck, M.; Faist, J.; Liu, H.C.; Sirtori, C. Injection locking of mid-infrared quantum cascade laser at 14 GHz, by direct microwave modulation. *Laser Photonics Rev.* **2014**, *8*, 443–449. [CrossRef]
24. Mottaghizadeh, A.; Asghari, Z.; Amanti, M.; Gacemi, D.; Vasanelli, A.; Sirtori, C. Ultra-fast modulation of mid infrared buried heterostructure quantum cascade lasers. In Proceedings of the 42nd International Conference on Infrared, Millimeter, and Terahertz Waves (IRMMW-THz), Cancun, Mexico, 27 August–1 September 2017. [CrossRef]
25. iC-HN iCSY HN1M High-Speed Module. Available online: https://www.ichaus.de/upload/pdf/HN1M_evalmanual_A3en.pdf (accessed on 30 June 2021).
26. Pierściński, K.; Pierścińska, D.; Kuźmicz, A.; Sobczak, G.; Bugajski, M.; Gutowski, P.; Chmielewski, K. Coupled cavity Mid-IR quantum cascade lasers fabricated by dry etching. *Photonics* **2020**, *7*, 45. [CrossRef]
27. Electrical Model. Available online: https://www.alpeslasers.ch/?a=36,41 (accessed on 30 June 2021).
28. Yang, K.; Liu, J.; Zhai, S.; Zhang, J.; Zhuo, N.; Wang, L.; Liu, S.; Liu, F. Room-temperature quantum cascade laser packaged module at ~8 µm designed for high-frequency response. *Electron. Lett.* **2021**, 1–3. [CrossRef]
29. PCO-7120 Laser Diode Driver Module. Available online: https://ixapps.ixys.com/DataSheet/pco-7120_manual.pdf (accessed on 30 June 2021).
30. Ashok, P.; Ganesh, M. Optimum electrical pulse characteristics for efficient gain switching in QCL. *Optik* **2017**, *146*, 51–62. [CrossRef]
31. Hemingway, M. External Cavity Quantum Cascade Lasers. Ph.D. Thesis, University of Sheffield, Sheffield, UK, 2018.
32. Gwinner, S. How to Choose a Pulsed Laser Diode Driver. Available online: https://www.laserdiodecontrol.com/How-to-Choose-a-Pulsed-Laser-Diode-Driver (accessed on 30 June 2021).
33. Mikołajczyk, J. Data Link with a High-Power Pulsed Quantum Cascade Laser Operating at the Wavelength of 4.5 µm. *Sensors* **2021**, *21*, 3231. [CrossRef] [PubMed]

Review

Ammonia Gas Sensors: Comparison of Solid-State and Optical Methods

Zbigniew Bielecki [1,*], Tadeusz Stacewicz [2], Janusz Smulko [3] and Jacek Wojtas [1]

1. Institute of Optoelectronics, Military University of Technology, 00-908 Warsaw, Poland; jacek.wojtas@wat.edu.pl
2. Institute of Experimental Physics, Faculty of Physics, University of Warsaw, 02-093 Warsaw, Poland; tadeusz.stacewicz@fuw.edu.pl
3. Faculty of Electronics, Telecommunications and Informatics, Gdansk University of Technology, 80-233 Gdansk, Poland; janusz.smulko@pg.edu.pl
* Correspondence: zbigniew.bielecki@wat.edu.pl

Received: 19 June 2020; Accepted: 24 July 2020; Published: 25 July 2020

Abstract: High precision and fast measurement of gas concentrations is important for both understanding and monitoring various phenomena, from industrial and environmental to medical and scientific applications. This article deals with the recent progress in ammonia detection using in-situ solid-state and optical methods. Due to the continuous progress in material engineering and optoelectronic technologies, these methods are among the most perceptive because of their advantages in a specific application. We present the basics of each technique, their performance limits, and the possibility of further development. The practical implementations of representative examples are described in detail. Finally, we present a performance comparison of selected practical application, accumulating data reported over the preceding decade, and conclude from this comparison.

Keywords: ammonia detection; NH_3; MOX sensors; polymer sensors; laser absorption spectroscopy; CRDS; CEAS; MUPASS; PAS

1. Introduction

Ammonia is a highly toxic chemical substance, common in biological processes, and applied in technical installation processes (cooling systems, chemical industry, and motor vehicles). The American Conference of Industrial Hygienists has set a limit to ammonia concentration in air of 25 ppm for long-term exposure (8 h) and 35 ppm for short-term ones (15 min) [1]. In medicine, the concentration of ammonia in the breath between 2500 and 5000 ppb is directly related to organ dysfunction and diabetes [2]. Therefore, the design of novel techniques and sensors which allow accurate and fast in-situ detection of trace ammonia concentration is highly desirable. Such sensors should satisfy the specific requirements: high sensitivity, enhanced selectivity, short response time, reversibility, high reliability, low energy consumption, low cost, safety, broad range of measurement at various operation temperatures, etc.

The issue of ammonia detection by gas sensors has attracted many researchers. A recent review paper about ammonia sensing focused on chemical mechanisms of gas sensing by solid-state or electrochemical sensors, with limited details about optical methods [3]. In our studies, we focus on optical methods, developed in our research teams. Moreover, another way of solid-state sensors modulation by UV irradiation was proposed and discussed. This method was advanced in the research teams preparing this review. These problems have not been considered in such a way in other papers about gas sensing.

Very advanced approaches like gas chromatography–mass spectrometry (GC-MS) and selective ion flow tube–mass spectrometry (SIFT-MS) are accurate for NH_3 measurement, but their use is complicated and require qualified staff, laborious samples preparation procedures, and time consuming measurements, as well as not-compact and expensive instruments which are difficult to maintain. Many applications require faster and easier tools. Therefore, solid and optical detection methods are rapidly developing (Figure 1). In this paper, much attention was paid on the optical sensors.

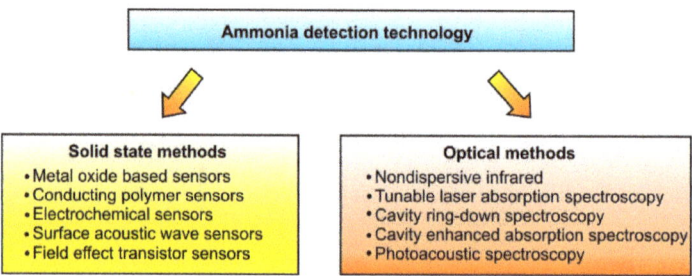

Figure 1. Ammonia detection technology.

We would like to underline that we preselected the considered ammonia gas sensors and there are other promising technologies which can be effectively used for ammonia sensing because of low production costs or measurement methods. Good examples are biosensors utilizing bacteria cultures or nanotechnology [4–8].

2. Solid-State Ammonia Sensors

Solid-state ammonia sensors can be divided into two groups considered in our paper: metal oxide-based sensors and conducting polymer sensors [3].

Metal oxide-based sensors (MOX) belong to the most investigated groups. Their main features offer simplicity, comprehensive detection action, miniature dimensions, flexibility in fabrication, long life expectancy, low cost, and serviceableness for alarm warning applications. MOX sensors change DC (static) resistance when exposed to ambient gas or humidity. Any change of MOX sensor resistance can trigger an alarm when toxic gas appears in an ambient atmosphere.

There is a variety of sensitive materials and methods of their preparation for NH_3 detection. Metal oxides, like SnO_2, ZnO, WO_3, TiO_2, and MoO_3, are most widely utilized for this purpose. They are classified into n-type and p-type [3]. Usually, the n-type semiconducting metal oxides are used for gas sensors due to their higher sensitivity [9]. The sensor comprises of grains of various diameter with a large active surface area. Smaller grains enable better sensing properties due to the larger ratio of active surface to the sensor volume. Atoms of oxygen are bound to the grains. When these atoms are displaced by other species present in ambient atmosphere, a potential barrier between the grains changes. As a result, the DC (static) resistance between the sensor's terminals changes.

Although the metal oxide sensors have attractive properties, their main disadvantage is a low selectivity in detecting of one particular component in a gas mixture [10]. Moreover, MOX sensors operate at elevated temperatures (up to a few hundreds of Celsius degrees), which requires additional energy for heating. Careful selection of ingredients, accelerating adsorption–desorption processes due to catalytic properties, helps to lower operating temperature and to enhance gas selectivity, pointing at selected gases. The sensor exhibits limited specificity for NH_3 and can be affected by acetone, ethanol, hydrogen, methane, nitric oxide, and nitrogen dioxide [11]. Artificial neural networks, conductance scanning at periodically varied temperature, as well as principle component or support vector machine analysis help to develop selective sensor systems, but only to some extend [12,13]. The algorithms support to determine even particular components of the gas mixtures. The average detection limit of these sensors ranges from 25 ppb to 100 ppm (Table 1). The selectivity can be further improved by noble metals doping [9,10] or selecting appropriate operating temperature [14]. Unfortunately, even these actions give limited results and require additional advances.

Another approach improving gas sensitivity and selectivity by MOX sensors utilize low-frequency noise (flicker noise) which can be more sensitive to the ambient atmosphere than DC resistance only [15,16]. This idea was proposed more than two decades ago and is still developed. DC resistance gives a single value due to a change of potential barrier at the presence of ambient gases. At the same time, the potential barrier fluctuates slowly, and these fluctuations are observed as flicker noise of the resistance fluctuations. Therefore, low-frequency noise measured as a power spectral density can be more informative than the value of DC resistance only. We confirmed experimentally that a single MOX sensor could detect two toxic gases, NH_3 and H_2S, by applying low-cost measurement set-up and flicker noise measurements [17]. MOX sensors are made of porous materials, which generate quite intense flicker noise components. Low-frequency noise is observed up to a few kHz at least and, therefore, can be measured using common electronic circuits (e.g., low-noise operational amplifiers and A/D converters sampling the signals up to tens of kHz only).

Some materials used for MOX sensors are photocatalytic (e.g., WO_3, TiO_2, Au nanoparticles functionalized with organic ligands). These materials can be modulated by UV-light irradiation (e.g., applying low-cost UV LEDs of different emitted wavelengths) [18]. UV light generates ions O_2^-, which are weakly bound to the surface of the grains and therefore enhance gas sensitivity. We confirmed experimentally that this modulation improves sensitivity at low gas concentrations and can be utilized for the detection of selected organic vapors (e.g., formaldehyde, NO_2 [18]).

The MOX sensors exhibit a temporal drift and ageing of their sensitivity. This detrimental effect can be reduced by algorithms of signal processing or by measuring the derivative parameters (e.g., change of DC resistance only) at relatively short time intervals [19,20]. Such effects are induced by the structure and technology of the MOX sensors, which comprise of grains of different size and morphology. Some ambient gases can be stably adsorbed by the grains and induce the ageing and drift of sensitivity. These effects can be reduced by pulse heating or intense UV irradiation, used for sensor cleansing [21].

Improvement in gas sensing stability can be reached by applying the materials of very repeatable structures. Two-dimensional materials (MoS_2, WS_2, phosphorene, graphene, carbon nanotubes) are of high interest for gas sensing due to their unique properties (high ratio of the sensing area to volume) and repeatable morphology (e.g., graphene layers) [22]. The sensors provide an opportunity to detect even a single gas molecule adsorbed by the active surface. An electronic device, using single-layer graphene for the gate of the field-effect transistor, can detect low gas concentrations and give repeatable results. Recent experimental studies confirmed that $1/f$ noise in such device has a component, called Lorentzian, of the frequency characteristic for the ambient gas (e.g., C_4H_8O, CH_3OH, C_2H_3N, $CHCl_3$) [23]. Moreover, the experiments gave repeatable results for different samples of electronic devices. We expect that such sensors should detect NH_3 molecules at very low concentrations in a similar way as the organic gases mentioned above. It was experimentally confirmed that the oxidized graphene sensor is suitable for NH_3 detection [24].

These impressive results were achieved for the structures, which are very repeatable and can be spoiled only by some limited impurities situated on the two-dimensional layers. Unfortunately, their production is expensive and requires specialized equipment. We may expect that low-cost MOX gas sensors might be produced using a mixture of graphene flakes decorated with the nanoparticles of the selected material for gas sensing. These sensors might be less sensitive than the presented electronic device. However, the morphology of repeatable two-dimensional structures should enhance their sensing properties in conjunction with very low costs and simplified technology of production. Further, gas sensing improvement can be achieved by UV-light irradiation because graphene is a photocatalytic material.

Flexible gas sensing materials are of great interest for emerging applications in wearable electronic devices for portable health monitoring applications. Detection of NH_3 is one of the hot topics in this area because ammonia may cause severe harm to the human body and is the most common air contaminant emitted from various sources (e.g., at the decomposition of protein products). The gas sensing materials should be transparent, mechanically durable, and operate at room temperature.

New structures of low-cost ammonia gas sensors were proposed recently [25–28]. These structures comprise different materials, including two-dimensional reduced graphene oxide [26], one-dimensional nanostructures (nanowires) [27], or polymers [25,28]. The sensors are chemo-resistive and, therefore, can be used in wearable or handheld portable applications. Moreover, their resistance can be monitored by utilizing triboelectric charging in self-powered wearable applications.

The second group of solid-state ammonia sensors consists of conducting polymer sensors. They represent important class of functional organic materials for next-generation sensors. Their features of high surface area, small dimensions, and unique properties have been used for various sensor constructions. Many remarkable examples have been reported over the past decade. The enhanced sensitivity of conducting polymer nanomaterials toward various chemical/biological species and external stimuli made them ideal candidates for incorporation into the design of the sensors. However, the selectivity and stability can be further improved.

Advances in nanotechnology allow the fabrication of various conducting polymer nanomaterials [29]. Among the conducting polymers, polyaniline (PANI), polypyrrole (PPy), polythiophene (PTh), and poly (butyl acrylate) (PBuA) or poly (vinylidene fluoride) (PVDF) are the most frequently used in the ammonia sensor as the active layer [30–32]. Gas sensors with conducting polymer are based on amperometric, conductometric, colorimetric, gravimetric, and potentiometric measuring techniques. However, their selectivity and stability should be also improved.

Some recently reported solid-state ammonia sensors are summarized in Table 1.

Table 1. Parameters of some solid-state ammonia sensors.

Sensor Material	Detection Limit	Response Time	Recovery Time	Operation Temperature	Reference
Metal Oxide					
SnO_2/In_2O_3	100 ppb	~0.1 min	10 s	RT	[33]
SnO_2/Pd/RGO	2 ppm	7 min	50 min	RT	[34]
SnO	5 ppm	<2 min	30 s	RT	[35]
TiO_2/GO/PANI	100 ppm	~0.5 min	17 s	RT	[36]
SnO_2 (type GGS10331)	5 ppm	<1 min	~few minutes	300–500 °C	[37]
Conducting Polymer					
PANI/SWNT	50 ppb	~few minutes	~few hours	RT	[38]
PANI-TiO_2-gold	10 ppm	-	-	RT	[39]
PANI/TiO_2	25 ppb	<1.5 min	-	RT	[40]
PANI/Cu	1 ppm	~0.1 min	160 s	RT	[41]
PANI/graphene	1 ppm	<1 min	23 s	RT	[42]

Where: GO—graphene oxide, RGO—reduced graphene oxide, RT—room temperature, PANI—polyaniline, PPy—polypyrrole, SWNT—single-walled carbon nanotubes.

Colorimetric sensors utilize optical measurements of the sample (porous matrix) interacting with the ambient gas. Therefore, we present these gas sensors in the group of solid-state sensors. They are used for the measurement of NH_3 in blood, urine, and wastewater. Colorimetric gas sensors are based on the change in color of a chemochromic reagent incorporated in a porous matrix such as porphyrin-based or pH indicator-based films. To detect this change in color, three basic components are needed: a light source, the chemochromic substance, and the light sensor. Liu et al. [43] has developed a solid-state, portable, and automated device capable to measure total ammonia amount in liquids, including the biological samples (e.g., urine). The idea of operation of the colorimetric sensor is shown in Figure 2. A horizontal gas flow channel passing through the sensing chamber, a red LED light source, and four photodiodes (a sensing-reference pair and a sensing-reference backup pair) were applied. The target gas is exposed to the sensor which then exhibits a color change proportional to the NH_3 concentration. The photodiodes convert the color change to electronic signals. Such a sensor is of high sensitivity, short response time, and fast reversibility for NH3 gas concentrations ranging from 2 ppm to 1000 ppm.

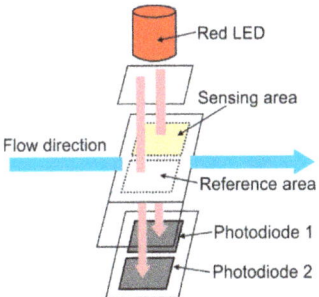

Figure 2. The schematic of the colorimetric optoelectronic ammonia sensor [43].

3. Optical Methods

Most of the optical approaches to ammonia detection are based on the effects of absorption and luminescence, more rarely on the refraction or the light reflection.

In an optical absorption technique, the measured gas is contained in the sensor chamber. The radiation passing through the chamber can be absorbed by gas molecules. Measuring the light absorption at the specific wavelength (λ), with possibly no other gas species absorbing in this spectral range, the concentration (N) of ammonia can be determined. The concentration estimation follows the Beer–Lambert absorption law

$$\alpha(\lambda)L = \sigma(\lambda)NL = ln\left(\frac{I_0}{I}\right) \quad (1)$$

where $\alpha(\lambda)$ is the absorption coefficient defined as the logarithmic ratio of the incident (I_0) and the transmitted (I) light intensities, L denotes the light path length in the chamber, and $\sigma(\lambda)$ denotes the absorption cross section. Ammonia exhibits UV and IR absorption bands around 135 nm, 155 nm, 195 nm, 1.5 µm, 2 µm, 3 µm, 4 µm, 6 µm, around 11 µm, and 16 µm (Figure 3). The cross-section (σ), defining the system sensitivity, is the highest in the UV region. It is about 10-times larger than at the 11 µm region (e.g., ~10^{-18} cm^2) and even 10^5-times larger than in other NIR ranges [44].

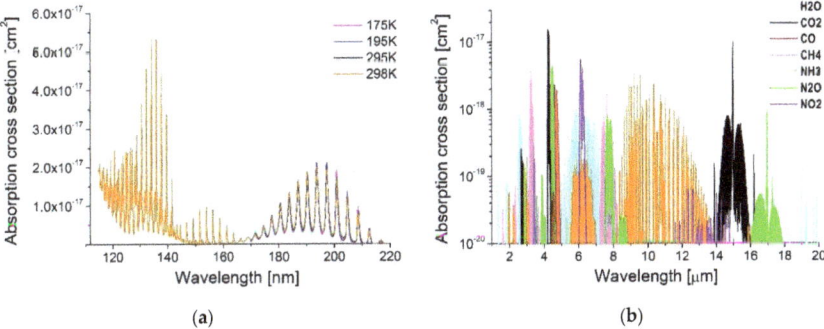

Figure 3. Absorption cross-section of ammonia for different temperatures in UV [45,46] (a). Absorption cross-section of selected molecules existing in standard conditions (1 atm, 288.2 K) calculated on the basis of the HITRAN database in IR range (b).

Unfortunately, in the UV and NIR range, there is a problem with interferences by water vapor and N_2O or NO_2 molecules, which occur at high concentrations in air. Their absorption bands might overlap with NH_3 spectra disturbing the measurements. Therefore, the selection of proper spectral range is crucial for successful optical detection.

Nondispersive infrared (NDIR) sensing belongs to simplest approaches of NH_3 detection. Figure 4 shows a schematic diagram of such a gas sensor. It usually consists of a broadband source (cheaper and smaller black body emitters or IR LEDs), absorption cell, optical filters, and detectors. Radiation from the broadband source passes through the chamber and two filters. The first filter covers the absorption band of the target gas (named active channel), while the other covers a non-absorbed spectral range (the reference channel). Transmission bands of the filters should not overlap with absorption bands of the other gases present in the chamber. Thus, absorption in the active channel is proportional to NH_3 concentration. The light transmitted through the reference channel is not attenuated. Therefore, it is used to compensate instabilities of the light source. The sensitivity of NDIR is influenced by the intensity of the source, the optical waveguide and detector parameters. The disadvantage of this technique is the low precision of the detecting small signal changes at an eventual large background, which results in low selectivity and a high detection limit.

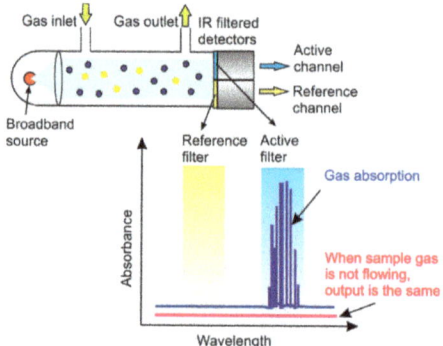

Figure 4. Design and principle of operation of the nondispersive infrared (NDIR) sensor.

There are various methods to improve the performance of NDIR sensors. For example, Max-IR Labs utilizes the NDIR technique together with fiber-optic evanescent wave spectroscopy (FEWS) for ammonia detection [47]. The IR radiation is transmitted through a silver-halide (AgClxBr1-x) optical waveguide without cladding and the detection performed by means of the evanescent field (Figure 5). The maximum of the peak due to ammonia absorption was observed at 1450 cm^{-1}. Such a sensor allows NH_3 detection beyond a 1 ppm limit.

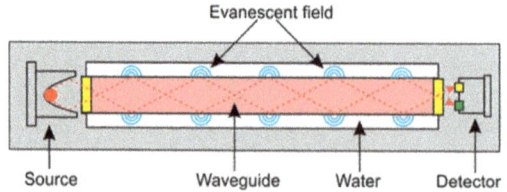

Figure 5. Diagram of the sensor based on NDIR and fiber-optic evanescent wave spectroscopy (FEWS) principles.

Laser spectroscopy is the best choice for trace gas analysis among optical approaches. It is characterized by high sensitivity and selectivity. Ammonia can be detected using tunable laser absorption spectroscopy (TLAS), multi-pass optical cell (MUPASS), cavity ring down spectroscopy (CRDS), cavity-enhanced absorption spectroscopy (CEAS), and photoacoustic (PAS) approach.

TLAS is a very interesting technique for ammonia concentration measuring using tunable diode lasers. The advantage of TLAS over other techniques consists in its ability to achieve low detection limits by using optical path extension techniques and improving the signal-to-noise ratio (SNR).

A single-pass TLAS system, operating in direct absorption measurement mode, consists of a tunable laser, transmitting optics, sample cell, receiving optics, photodetector, and a signal detection circuit (Figure 6). The optics matches the laser and photodetector to the gas inside the sample cell. A thermoelectric controller (TEC) is used to set the laser-operating temperature to a value where the desired wavelength can be reached due to injection current tuning. The injection current is usually scanned periodically with a ramp signal, which leads to laser wavelength scanning. The amplitude of the scan should cover the absorption transition of interest. Laser radiation passes through the absorption cell and the transmitted signal is measured using the photodetector. In absence of the absorption, the detector signal represents laser power changes vs. the current. When absorption occurs, a dip in transmission is observed. Ratio of the signals registered at the line center corresponding to the case without absorption (I_0) and with absorption (I_1) can be used to calculate the gas concentration according to the formula (1).

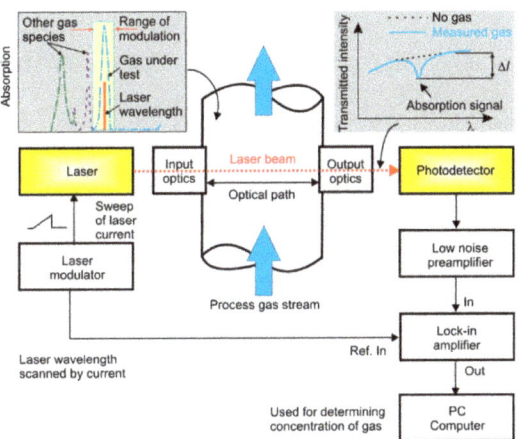

Figure 6. Block diagram and principle of operation of the typical tunable laser absorption spectroscopy (TLAS) sensor.

Based on TLAS, different components such as NH_3, CO, O_2, CH_4, H_2O, CO_2, and HCl can be detected with high selectivity and sensitivity. TLAS has been employed for various applications, including industrial process monitoring and its control, environmental monitoring, combustion and flow analyses, trace species measurements, and so on. However, the sensitivity of this technique is usually limited to the absorption coefficient value of $\sim 10^{-2}$–10^{-3} cm^{-1} (Equation (1)).

Wavelength modulation spectroscopy (WMS) is a useful technique providing SNR improvement, which is used in high sensitivity applications [48]. Small modulation (with frequency $f \approx 1$–20 kHz) is added to the laser current scan, which is provided in a similar way as in TLAS (Figure 7). The transmitted signal, detected by the photodiode, is fed to a lock-in amplifier. As the laser is scanned across an absorption line profile, the transmitted on-absorption signal changes at the frequency of $2f$, while the transmitted off-absorption signal changes at the frequency of f (Figure 8).

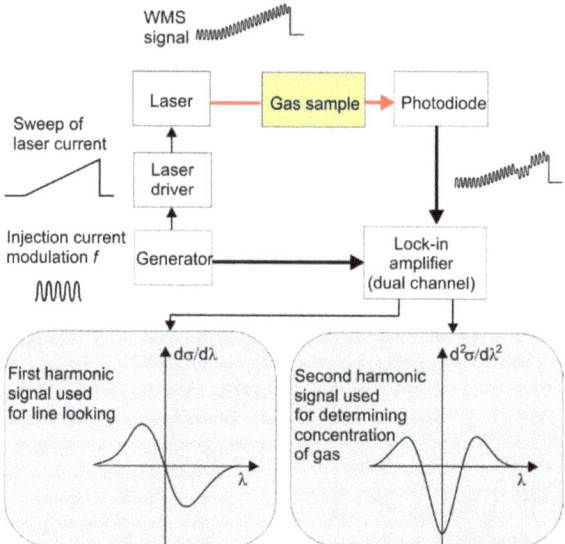

Figure 7. The principle of operation of the sensor using a combination of TLAS and wavelength modulation spectroscopy (WMS) techniques.

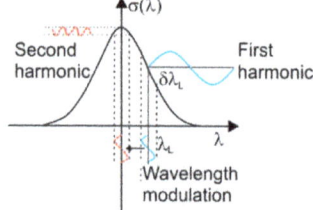

Figure 8. Idea of first and second harmonic detection with tunable diode lasers.

Application of high frequency minimizes $1/f$ noise. Therefore, setting the lock-in band around second harmonic ensures that the sensor becomes more suitable for high sensitivity applications compared to the standard TLAS approach. Typical sensitivity limits the absorption coefficient achievable with WMS and is about 10^{-4} cm^{-1}. Better limits, about 10^{-5}–10^{-7} cm^{-1}, can be obtained for balanced detection-based WMS [49]. This method involves an electrical circuit that subtracts the photocurrents of two detectors: one of them registers the reference laser intensity while the other measures the signal passing through the absorption cell. That provides opportunity to reject common mode laser noise.

Small absorptions, which occur due to low densities of molecules in the samples or due to weak line strengths, are usually compensated by extending the optical path with multi-pass cells (White or Herriott) or cavity-enhanced methods. They enable the optical path to be extended. If instead of a single-pass chamber we apply the multi-pass cell (Figure 9), the sensitivity of the sensor would be improved by the factor of the optical path lengthening. A key parameter of multi-pass cells is the ratio of path length to volume. Moreover, this system provides an opportunity for the simultaneous application of the WMS approach and *2f* detection.

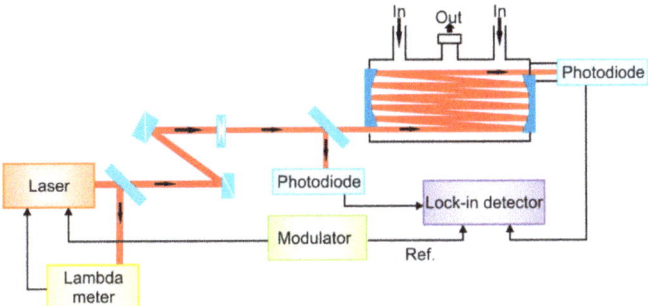

Figure 9. Scheme of a multi-pass experimental system for multi-pass optical cell (MUPASS)–WMS spectroscopy.

Unlike conventional configurations, which involve at least a pair of mirrors separated by exactly defined distances, circular multi-pass cells have been developed (Figure 10) [50]. Thanks to the single piece, the cell is especially robust against thermal expansion. Minimizing the cell size is desired for the development of fast and portable gas sensors.

Figure 10. The circular multi-pass—small volume cell.

Cavity ring-down spectroscopy (CRDS) exploits the sample cells in the form of optical cavities (resonators) built with mirrors of very high reflectivity. A simplified scheme of such experiment is presented in Figure 11. Laser pulse injected into the cavity through one of the mirrors is then reflected many times among them. Its wavelength must be tuned to the NH_3 spectral line in order to measure the absorption coefficient of ammonia contained inside. The light transmitted through the exit mirror is monitored by a detector. Analysis of the output signal by the acquisition system provides opportunity to determine the Q-factor of the cavity which is limited due to diffraction and mirror losses as well as due to the light absorption or scattering inside the cell. As far as the Q-factor is inversely proportional to the signal decay time, its value is found due to analysis of the photoreceiver signal by the acquisition system. The majority of the approaches exploit a two-step procedure consisting of the decay time measurement: first, when the cavity is empty (τ_0), and then, when the cavity is filled with the tested gas (τ_A). These decay times depend on the mirror reflectivity, resonator length, and extinction factor (absorption and scattering of light in the cavity) [51]. Comparing both decay times, the absorber concentration can be found.

$$N = \frac{1}{c\sigma(\lambda)} \left(\frac{1}{\tau_A} - \frac{1}{\tau_0} \right) \qquad (2)$$

where c is the light speed and $\sigma(\lambda)$ is the absorption cross-section.

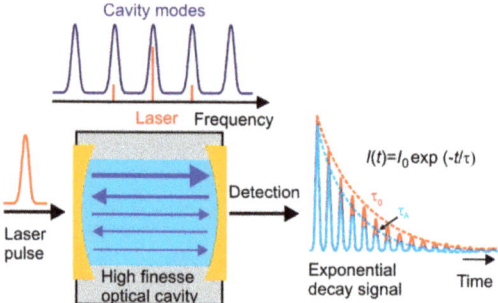

Figure 11. Idea of the cavity ring down spectroscopy (CRDS).

There are various approaches to CRDS with pulsed lasers or with continuously operating AM modulated ones. Using these techniques, the absorption coefficients $\alpha = \sigma N < 10^{-9}$ cm^{-1} can be observed. The detection limit is mainly related to the resonator quality (determined by τ_0), but also by the precision of decay time measurement. The advantage of CRDS over other absorption spectroscopy approaches consists not only in dominant sensitivity. Their superiority also results from minimized impacts of light source intensity fluctuations or detector sensitivity changes on the measurement results.

This method is extremely sensitive but requires that the laser frequency is precisely matched to the cavity mode. In this way, it is possible to obtain a high Q-factor of the resonator and efficient storage of optical radiation. On the other hand, small mechanical instabilities cause changes in the cavity mode frequency and significant output signal fluctuations, which is the main disadvantage of this method [52]. It can be minimalized by use of cavity length stabilization [53,54], or by application of the cavity with dense mode structure, called cavity-enhanced absorption spectroscopy (CEAS) [55].

In CEAS, the off-axis arrangement of the laser and cavity is applied. Similarly, to the conventional CRDS system, the light is repeatedly reflected by the mirrors. However, the reflected beams, which correspond to different trips, are spatially separated inside the resonator. The use of highly reflecting mirrors provides huge extension of the effective optical path. As result, either a dense mode structure of low finesse occurs, or the mode structure does not establish at all. In this way, sharp resonances of the cavity are avoided, so the system is much less sensitive to mechanical instabilities. The CEAS sensors attain the detection limit of about 10^{-9} cm^{-1} [56]. Detailed information about CRDS and CEAS techniques is presented in the authors' papers [57–59].

Among the so-called in-situ sensors, photoacoustic spectroscopy (PAS) belongs to the most popular one. In PAS, conversion of laser light energy into an acoustic wave is applied. The gas sample, placed in photoacoustic chamber, is irradiated by optical radiation, AM modulated with acoustic frequency. The laser wavelength is matched to the absorption line of a molecule of interest. If the absorber is present in the cell, a portion of optical radiation is converted onto heat energy. Then, a local and periodic growth of temperature and pressure occurs (Figure 12). The resulting acoustic wave is detected at modulation frequency by a very sensitive microphone placed in the chamber. The PAS signal, which is proportional to the absorber concentration (N), is given by

$$A(T,\lambda) \propto P_o N\alpha(T,\lambda) L \frac{Q}{fV}\eta = P_o N\alpha(T,\lambda)\frac{Q}{fA}\eta \qquad (3)$$

where P_o denotes average laser power, Q is the quality factor of the resonant cell, f is the frequency of modulation, V is the gas volume, A is the cross-sectional area, and η is the system efficiency factor (e.g., microphone efficiency and loss factors).

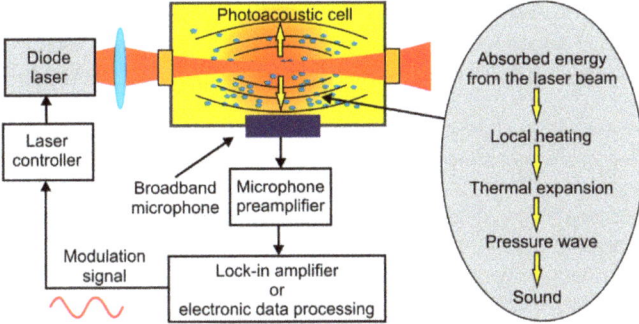

Figure 12. Idea of photoacoustic spectroscopy.

The microphone output signal is recorded using a low-noise preamplifier and a lock-in voltmeter. The strongest effect is achieved when the modulation frequency is matched to the resonance of the photoacoustic chamber. In order to increase the signal, various constructions of photoacoustic cells (acoustic resonators with higher Q-factor) are applied.

The improvement of this technique, which allows to achieve very high detection sensitivity, is the quartz-enhanced photoacoustic spectroscopy (QEPAS). The basic principle of operation is similar, but here the laser beam is focused between the U-shaped prongs of a resonant piezo-quartz fork [60]. This type of acoustic transducer is sensitive to signals generated by asymmetric prongs oscillations caused by acoustic wave, induced by the AM-modulated laser radiation. External sources of interference do not provide output signals because they cause symmetrical oscillations. Similarly to the conventional PAS system, in QEPAS set-ups, the measurements are performed with a wavelength modulation technique using the $2f$ detection [61]. Combined with the high Q-factor of the quartz fork, it gives the opportunity to build ultra-compact sensors that reach detection limits comparable to that of CEAS. In case of ammonia, it corresponds to a few ppb or even sub-ppb level.

Recently reported parameters of some optical ammonia sensors are given in Table 2.

Table 2. Parameters of some optical ammonia sensors.

Sensing Methods	Detection Limit	Radiation Source	Wavelength	Other Parameter	Reference
NDIR	1 ppm	Deuterium lamp	Filter 200–225 nm	λ_{center} = 205 nm FWHM = 10 nm	[44]
MUPASS-WMS	7 ppb	QCL	1103.44 cm^{-1}	Eff. path length—76.45 m	[62]
CRDS-cw (open-path)	1.3 ppb	QCL-DFB	10.33 µm	R = 0.9995, L = 50 cm	[63]
CRDS	0.74 ppb	QCL (cw-EC)	6.2 µm	R = 0.9998, L = 50 cm, p = 115 Torr	[64]
CEAS	15 ppb	QCL-DFB (pulsed)	967.35 cm^{-1}	L = 53 cm, t_i = 5–10 ns	[65]
PAS	0.7 ppb	EC-QCL	10.36 µm	p = 220 Torr	[66]
QEPAS	<10 ppb	QCL (cw-EC)	10.6 µm	Ammonia detection in exhaled breath	[67]

QCL—quantum cascade laser, PAS—photoacoustic spectroscopy, CRDS—cavity ring down spectroscopy, QEPAS—quartz-enhanced photoacoustic spectroscopy, WMS—wavelength modulation spectroscopy, R—reflectivity of mirrors, L—optical cavity length, EC—external cavity, p—pressure. Other methods of gas detection were described in our earlier studies [68,69].

Among many applications, the ammonia sensors for medical purpose are an important part of such detectors. The normal concentration of ammonia in the breath of a healthy man is in the range of 0.25–2.9 ppm [70]. An excessive concentration might suggest renal failure, *Helicobacter Pyroli*, diabetes, and oral cavity disease [71,72]. The main problem in the construction of ammonia sensors for breath analysis does not consist in extreme sensitivity but in high selectivity. More than 3000 various constituents were already detected in exhaled air [73]. Their absorption spectra might overlap the spectral fingerprint of the ammonia and disturb the measurement. Water vapor and carbon dioxide are the main interferences since their concentrations can reach up to 5% in breath, and thus it exceeds the ammonia density by many orders of magnitude. Carbon monoxide and methane should also be taken into account.

As it was mentioned above, the highest sensitivities of ammonia detection using laser absorption spectroscopy can be achieved in UV and in a range of 10–11 µm, due to the largest values of the absorption cross-sections. However, the detection at the wavelengths used in telecommunication 1.4–1.6 µm is more convenient because of a large variety of relatively cheap laser sources, optical elements, and photodetectors. The measurements of ammonia around 1.51 µm with the detection limit of 4 ppm was already demonstrated using the CEAS technique [74]. We found that suitable wavelengths for ammonia detection are also the lines at 1.5270005 µm and 1.5270409 µm (Figure 13). The single-mode laser wavelength was controlled with the HighPrecision lambdameter (model WS6) ensuring precision in a short time (1 min) of 0.0005 nm. The NH_3 absorption coefficient (at 2 ppm) reaches here about $3.5 \cdot 10^{-6}$ cm^{-1}, and experimental techniques like MUPASS–WMS can be effectively used. In this range, carbon dioxide and water vapor interference is about eight times weaker.

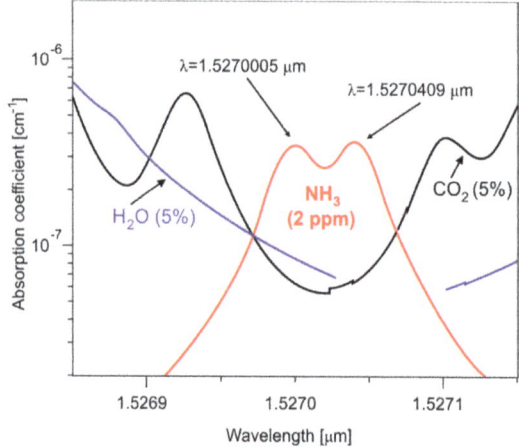

Figure 13. Absorption spectra of ammonia, H_2O, and CO_2 around 1.527 µm [75].

Our ammonia sensor used the MUPASS–WMS approach (Figure 9). A single-mode diode laser (Toptica, model DL100, 20 mW) was applied as a light source. Output signals were integrated over 1 min. Results of sensor investigation are presented in Figure 14. It provides proper results for NH_3 concentrations of up to about 1 ppm. For lower values, the deviation from a linear characteristic is observed. This is probably a result of poor regulation of reference concentration inside the sensor due to NH_3 deposit on the walls of the system, which causes a systematic error of measured data. In order to avoid the concentration measurement disturbed by NH_3 molecules adsorbed on the walls (and then desorbed), the sensor was kept at a temperature of 50 °C.

Nevertheless, due to the proper choice of spectral lines, we achieved a good immunity of the detection against H_2O and CO_2 at the concentrations which might occur in breath (Figure 13). The detection limit of 1 ppm was better than in other experiments preformed at a spectral range of 1.5 µm [74]. Our multi-pass sensor is suitable for rough monitoring of ammonia in the exhaled air and for detection of morbid states (>1.5 ppm).

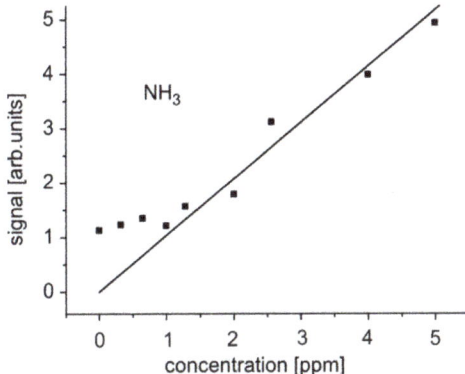

Figure 14. Results of ammonia measurement with MUPASS–WMS system [75].

4. Conclusions

Based on the analysis, it can be concluded that the further development of in-situ ammonia detection technologies will focus on the improvement of sophisticated laboratory equipment, e.g., the mass spectrometers and lasers. One can expect the development of sensors dedicated to a specific application. This article focuses on the second group of the sensors in which two technologies with high potential for application and development have been distinguished: solid-state ammonia sensors and laser absorption spectroscopy. Metrological and operational parameters of such sensors largely depend on the application. Basically, they must be characterized by fast operation, the largest measuring range and the highest measurement precision, high reliability, selectivity, as well as small dimensions, and low cost. Sensors belonging to the former group can detect the ammonia with a concentration of several ppm within tens of seconds. Their advantage consists in very small dimensions and very low cost. Materials engineering and nanotechnology are key technologies for the development of this sensor group. We underlined that recent advances in materials and measurement methods enhance gas detection by such low-cost gas sensors in portable applications. We are conscious that our research presents a limited number of NH_3 gas sensors. This issue is a hot topic in the gas sensing area, and other attractive sensors were developed recently (e.g., capacitive sensors [76], colorimetric sensing materials [77], or advanced MOX sensors [78]). The latter group of sensors can detect up to four orders of lower ammonia in fractions of a second. However, their costs are much higher, mainly due to the lasers and high-quality optical components. The further improvement of these sensors is much more dependent on the development of lasers [79] than photodetectors [80], because they are already at a very high level. It mainly concerns a progress in lasers for mid-infrared radiation, since one can expect the highest sensitivity and selectivity in this spectral range. However, miniaturization, reliability, and reduction of production costs are still a current challenge for all optical sensor components. It should be also mentioned that MOX sensors can utilize optical methods in some sense by applying UV light to improve the sensitivity and selectivity of gas detection. Therefore, both types of gas sensors start to interlace MOX and optical sensors technologies to reduce costs and popularize their applications. We believe that different types of gas sensors find complementary areas of new applications.

Author Contributions: Z.B.: conceptualization, writing—original draft, visualization. T.S.: writing—review & editing, investigation. J.S.: writing—review & editing, investigation. J.W.: formal analysis, investigation, writing—review & editing. All authors have read and agreed to the published version of the manuscript.

Funding: The publication was supported by The Polish National Centre for Research and Development as part of the "Sense" project, ID 347510.

Conflicts of Interest: The authors declare no conflict of interest.

References

1. Manap, H.; Mazlee, N.N.; Suzalina, K.; Najib, M.S. An open-path optical fibre sensor for ammonia measurement in the ultraviolet region. *ARPN J. Eng. Appl. Sci.* **2016**, *11*, 10940–10943.
2. Solga, S.F.; Mudalel, M.L.; Spacek, L.A.; Risby, T.H. Fast and accurate exhaled breath ammonia measurement. *J. Vis. Exp.* **2014**, *88*, 51658. [CrossRef]
3. Kwak, D.; Lei, Y.; Maric, R. Ammonia gas sensors: A comprehensive review. *Talanta* **2019**, *204*, 713–730. [CrossRef] [PubMed]
4. Zhang, Q.; Ding, J.; Kou, L.; Kui, W. A potentiometric flow biosensor based on ammonia-oxidizing bacteria for the detection of toxicity in water. *Sensors* **2013**, *13*, 6936–6945. [CrossRef]
5. Bollmann, A.; Revsbech, N.P. An NH_4^+ biosensor based on ammonia-oxidizing bacteria for use under anoxic conditions. *Sens. Actuators B Chem.* **2005**, *105*, 412–418. [CrossRef]
6. Bianchi, R.C.; Rodrigo da Silva, E.; Dall'Antonia, L.H.; Ferreira, F.F.; Alves, W.A. A nonenzymatic biosensor based on gold electrodes modified with peptide self-assemblies for detecting ammonia and urea oxidation. *Langmuir* **2014**, *30*, 11464–11473. [CrossRef]
7. Khan, I.R.; Mohammad, A.; Asiri, A.M. (Eds.) *Advanced Biosensors for Health Care Application*, 1st ed.; Elsevier: Amsterdam, The Netherlands; Cambridge, UK; Oxford, UK, 2019. [CrossRef]
8. Nasiri, N.; Clarke, C. Nanostructured gas sensors for medical and health applications: Low to high dimensional materials. *Biosensors* **2019**, *9*, 43. [CrossRef]
9. Kim, H.J.; Lee, J.H. Highly sensitive and selective gas sensors using p-type oxide semiconductors: Overview. *Sens. Actuators B Chem.* **2014**, *192*, 607–627. [CrossRef]
10. Srivastava, V.; Jain, K. Highly sensitive NH_3 sensor using Pt catalyzed silica coating over WO_3 thick films. *Sens. Actuators B Chem.* **2008**, *133*, 46–52. [CrossRef]
11. Georges, J. Determination of ammonia and urea in urine and of urea in blood by use of an ammonia-selective electrode. *Clin. Chem.* **1979**, *25*, 1888–1890. [CrossRef]
12. Timmer, B.; Olthuis, W.; Van Den Berg, A. Ammonia sensors and their applications—A review. *Sens. Actuators B Chem.* **2005**, *107*, 666–677. [CrossRef]
13. Lentka, Ł.; Smulko, J.M.; Ionescu, R.; Granqvist, C.G.; Kish, L.B. Determination Of Gas Mixture Components Using Fluctuation Enhanced Sensing And The LS-SVM Regression Algorithm. *Metrol. Meas. Syst.* **2015**, *22*, 341–350. [CrossRef]
14. Korotcenkov, G.; Cho, B.K. Engineering approaches for the improvement of conductometric gas sensor parameters: Part 1. Improvement of sensor sensitivity and selectivity (short survey). *Sens. Actuators B Chem.* **2013**, *188*, 709–728. [CrossRef]
15. Kish, L.B.; Vajtai, R.; Granqvist, C.G. Extracting information from noise spectra of chemical sensors: Single sensor electronic noses and tongues. *Sens. Actuators B Chem.* **2000**, *71*, 55–59. [CrossRef]
16. Dziedzic, A.; Kolek, A.; Licznerski, B. Noise and nonlinearity of gas sensors–preliminary results. In Proceedings of the 22nd International Spring Seminar on Electronics Technology, Dresden-Freital, Germany, 18–20 May 1999; pp. 99–104.
17. Kotarski, M.; Smulko, J. Hazardous gases detection by fluctuation-enhanced gas sensing. *Fluct. Noise Lett.* **2010**, *9*, 359–371. [CrossRef]
18. Trawka, M.; Smulko, J.; Hasse, L.; Granqvist, C.G.; Annanouch, F.E.; Ionescu, R. Fluctuation enhanced gas sensing with WO_3-based nanoparticle gas sensors modulated by UV light at selected wavelengths. *Sens. Actuators B Chem.* **2016**, *234*, 453–461. [CrossRef]
19. Kwiatkowski, A.; Chludziński, T.; Smulko, J. Portable exhaled breath analyzer employing fluctuation-enhanced gas sensing method in resistive gas sensors. *Metrol. Meas. Syst.* **2018**, *25*, 551–560.
20. Gutierrez-Osuna, R. Pattern analysis for machine olfaction: A review. *IEEE Sens. J.* **2002**, *2*, 189–202. [CrossRef]
21. Balandin, A.A.; Rumyantsev, S. Low-Frequency Noise in Low-Dimensional van der Waals Materials. *arXiv* **2019**, arXiv:1908.06204.
22. Donarelli, M.; Ottaviano, L. 2d materials for gas sensing applications: A review on graphene oxide, MoS_2, WS_2 and phosphorene. *Sensors* **2018**, *18*, 3638. [CrossRef]
23. Rumyantsev, S.; Liu, G.; Potyrailo, R.A.; Balandin, A.A.; Shur, M.S. Selective sensing of individual gases using graphene devices. *IEEE Sens. J.* **2013**, *13*, 2818–2822. [CrossRef]

24. Bannov, A.G.; Prášek, J.; Jašek, O.; Shibaev, A.A.; Zajíčková, L. Investigation of Ammonia Gas Sensing Properties of Graphite Oxide. *Procedia Eng.* **2016**, *168*, 231–234. [CrossRef]
25. Cai, J.; Zhang, C.; Khan, A.; Liang, C.; Li, W.D. Highly transparent and flexible polyaniline mesh sensor for chemiresistive sensing of ammonia gas. *RSC Adv.* **2018**, *8*, 5312–5320. [CrossRef]
26. Duy, L.T.; Trung, T.Q.; Dang, V.Q.; Hwang, B.U.; Siddiqui, S.I.Y.; Yoon, S.K.; Chung, D.Y.; Lee, N.E. Flexible transparent reduced graphene oxide sensor coupled with organic dye molecules for rapid dual-mode ammonia gas detection. *Adv. Funct. Mater.* **2016**, *26*, 4329–4338. [CrossRef]
27. Tang, N.; Zhou, C.; Xu, L.; Jiang, Y.; Qu, H.; Duan, X. A Fully Integrated Wireless Flexible Ammonia Sensor Fabricated by Soft Nano-Lithography. *ACS Sens.* **2019**, *4*, 726–732. [CrossRef]
28. Kumar, L.; Rawal, I.; Annapoorni, S. Flexible room temperature ammonia sensor based on polyaniline. *Sens. Actuators B Chem.* **2017**, *240*, 408–416. [CrossRef]
29. Yoon, H. Current Trends in Sensors Based on Conducting Polymer Nanomaterials. *Nanomaterials* **2013**, *3*, 524–549. [CrossRef]
30. Sonkusare, G.; Tyagi, S.; Kumar, R.; Mishra, S.; Author, C. Room Temperature Ammonia Gas Sensing Using Polyaniline Nanoparticles Based Sensor. *Int. J. Mater. Sci.* **2017**, *12*, 283–291. [CrossRef]
31. Šetka, M.; Drbohlavová, J.; Hubálek, J. Nanostructured polypyrrole-based ammonia and volatile organic compound sensors. *Sensors* **2017**, *17*, 562. [CrossRef]
32. Bai, H.; Shi, G. Gas sensors based on conducting polymers. *Sensors* **2007**, *7*, 267–307. [CrossRef]
33. Qi, Q.; Wang, P.; Zhao, J.; Feng, L.; Zhou, L.; Xuan, R.; Liu, Y.; Li, G. SnO_2 nanoparticle-coated In_2O_3 nanofibers with improved NH_3 sensing properties. *Sens. Actuators B Chem.* **2014**, *194*, 440–446. [CrossRef]
34. Su, P.G.; Yang, L.Y. NH_3 gas sensor based on $Pd/SnO_2/RGO$ ternary composite operated at room-temperature. *Sens. Actuators B Chem.* **2016**, *223*, 202–208. [CrossRef]
35. Wu, H.; Ma, Z.; Lin, Z.; Song, H.; Yan, S.; Shi, Y. High-sensitive ammonia sensors based on tin monoxide nanoshells. *Nanomaterials* **2019**, *9*, 388. [CrossRef] [PubMed]
36. Tian, J.; Yang, G.; Jiang, D.; Su, F.; Zhang, Z. A hybrid material consisting of bulk-reduced TiO_2, graphene oxide and polyaniline for resistance based sensing of gaseous ammonia at room temperature. *Microchim. Acta* **2016**, *183*, 2871–2878. [CrossRef]
37. Wozniak, L.; Kalinowski, P.; Jasinski, G.; Jasinski, P. FFT analysis of temperature modulated semiconductor gas sensor response for the prediction of ammonia concentration under humidity interference. *Microelectron. Reliab.* **2018**, *84*, 163–169. [CrossRef]
38. Zhang, T.; Nix, M.B.; Yoo, B.Y.; Deshusses, M.A.; Myung, N.V. Electrochemically functionalized single-walled carbon nanotube gas sensor. *Electroanalysis* **2006**, *18*, 1153–1158. [CrossRef]
39. Liu, C.; Tai, H.; Zhang, P.; Ye, Z.; Su, Y.; Jiang, Y. Enhanced ammonia-sensing properties of $PANI-TiO_2$-Au ternary self-assembly nanocomposite thin film at room temperature. *Sens. Actuators B Chem.* **2017**, *246*, 85–95. [CrossRef]
40. Li, Y.; Gong, J.; He, G.; Deng, Y. Fabrication of polyaniline/titanium dioxide composite nanofibers for gas sensing application. *Mater. Chem. Phys.* **2011**, *129*, 477–482. [CrossRef]
41. Patil, U.V.; Ramgir, N.S.; Karmakar, N.; Bhogale, A.; Debnath, A.K.; Aswal, D.K.; Gupta, S.K.; Kothari, D.C. Room temperature ammonia sensor based on copper nanoparticle intercalated polyaniline nanocomposite thin films. *Appl. Surf. Sci.* **2015**, *339*, 69–74. [CrossRef]
42. Wu, Z.; Chen, X.; Zhu, S.; Zhou, Z.; Yao, Y.; Quan, W.; Liu, B. Enhanced sensitivity of ammonia sensor using graphene/polyaniline nanocomposite. *Sens. Actuators B Chem.* **2013**, *178*, 485–493. [CrossRef]
43. Liu, N.Y.; Cay-Durgun, P.; Lai, T.; Sprowls, M.; Thomas, L.; Lind, M.L.; Forzani, E.A. Handheld, Colorimetric Optoelectronic Dynamics Analyzer for Measuring Total Ammonia of Biological Samples. *IEEE J. Transl. Eng. Health Med.* **2018**, *6*, 1–10. [CrossRef] [PubMed]
44. Fischbacher, B.; Lechner, B.; Brandstätter, B. Ammonia distribution measurement on a hot gas test bench applying tomographical optical methods. *Sensors* **2019**, *19*, 896. [CrossRef] [PubMed]
45. Limão-Vieira, P.; Jones, N.C.; Hoffmann, S.V.; Duflot, D.; Mendes, M.; Lozano, A.I.; Ferreira da Silva, F.; García, G.; Hoshino, M.; Tanaka, H. Revisiting the photoabsorption spectrum of NH_3 in the 5.4–10.8 eV energy region. *J. Chem. Phys.* **2019**, *151*, 184302. [CrossRef]
46. Chen, F.; Judge, D.; Wu, C.Y.R.; Caldwell, J. Low and room temperature photoabsorption cross sections of NH_3 in the UV region. *Planet. Space Sci.* **1998**, *47*, 261–266. [CrossRef]

47. Roodenko, K.; Hinojos, D.; Hodges, K.L.; Veyan, J.F.; Chabal, Y.J.; Clark, K.; Katzir, A.; Robbins, D. Non-dispersive infrared (NDIR) sensor for real-time nitrate monitoring in wastewater treatment. In Proceedings of the Optical Fibers and Sensors for Medical Diagnostics and Treatment Applications XIX, San Francisco, CA, USA, 2–3 February 2019; Volume 15, p. 10872. [CrossRef]
48. Klein, A.; Witzel, O.; Ebert, V. Rapid, Time-Division multiplexed, Direct Absorptionand wavelength modulation-spectroscopy. *Sensors* **2014**, *14*, 21497–21513. [CrossRef] [PubMed]
49. Mitra, C.; Sharma, R. Diode Laser-Based Sensors for Extreme Harsh Environment Data Acquisition. In *High Energy and Short Pulse Lasers*; InTech: London, UK, 2016; p. 13. [CrossRef]
50. IRcell–Long Path in a Compact Design. Available online: https://dea47l1p89u26.cloudfront.net/wp-content/uploads/2017/12/IRcell-green-product.jpg (accessed on 20 February 2020).
51. O'neill, H.; Gordon, S.; O'Neill, M.; Gibbons, R.; Szidon, J. A computerized classification technique for screening for the presence of breath biomarkers in lung cancer. *Clin. Chem.* **1988**, *34*, 1613–1618. [CrossRef] [PubMed]
52. Berden, G.; Peeters, R.; Meijer, G. Cavity ring-down spectroscopy: Experimental schemes and applications. *Int. Rev. Phys. Chem.* **2010**, *19*, 565–607. [CrossRef]
53. Cygan, A.; Lisak, D.; Masłowski, P.; Bielska, K.; Wójtewicz, S.; Domysławska, J.; Trawiński, R.S.; Ciuryło, R.; Abe, H.; Hodges, J.T. Pound-Drever-Hall-locked, frequency-stabilized cavity ring-down spectrometer. *Rev. Sci. Instrum.* **2011**, *82*, 063107. [CrossRef]
54. Cygan, A.; Wójtewicz, S.; Domysławska, J.; Masłowski, P.; Bielska, K.; Piwiński, M.; Stec, K.; Trawiński, R.S.; Ozimek, F.; Radzewicz, C.; et al. Spectral line-shapes investigation with Pound-Drever-Hall-locked frequency-stabilized cavity ring-down spectroscopy. *Eur. Phys. J. Spec. Top.* **2013**, *222*, 2119–2142. [CrossRef]
55. Engeln, R.; Berden, G.; Peeters, R.; Meijer, G. Cavity enhanced absorption and cavity enhanced magnetic rotation spectroscopy. *Rev. Sci. Instrum.* **1998**, *69*, 3763–3769. [CrossRef]
56. Wojtas, J.; Czyzewski, A.; Stacewicz, T.; Bielecki, Z. Sensitive detection of NO_2 with cavity enhanced spectroscopy. *Opt. Appl.* **2006**, *36*, 461–467.
57. Wojtas, J.; Mikolajczyk, J.; Nowakowski, M.; Rutecka, B.; Medrzycki, R.; Bielecki, Z. Applying CEAS method to UV, VIS, and IR spectroscopy sensors. *Bull. Pol. Acad. Sci. Tech. Sci.* **2011**, *59*, 415–418. [CrossRef]
58. Stacewicz, T.; Wojtas, J.; Bielecki, Z.; Nowakowski, M.; Mikołajczyk, J.; Mędrzycki, R.; Rutecka, B. Cavity ring down spectroscopy: Detection of trace amounts of substance. *Opto-Electron. Rev.* **2012**, *20*. [CrossRef]
59. Wojtas, J.; Bielecki, Z.; Stacewicz, T.; Mikołajczyk, J.; Nowakowski, M. Ultrasensitive laser spectroscopy for breath analysis. *Opto-Electron. Rev.* **2012**, *20*. [CrossRef]
60. Zheng, H.; Liu, Y.; Lin, H.; Liu, B.; Gu, X.; Li, D.; Huang, B.; Wu, Y.; Dong, L.; Zhu, W.; et al. Quartz-enhanced photoacoustic spectroscopy employing pilot line manufactured custom tuning forks. *Photoacoustics* **2020**, *17*, 100158. [CrossRef]
61. Hu, L.; Zheng, C.; Zheng, J.; Wang, Y.; Tittel, F.K. Quartz tuning fork embedded off-beam quartz-enhanced photoacoustic spectroscopy. *Opt. Lett.* **2019**, *44*, 2562–2565. [CrossRef]
62. Owen, K.; Farooq, A. A calibration-free ammonia breath sensor using a quantum cascade laser with WMS 2f/1f. *Appl. Phys. B Lasers Opt.* **2014**, *116*, 371–383. [CrossRef]
63. Shadman, S.; Rose, C.; Yalin, A.P. Open-path cavity ring-down spectroscopy sensor for atmospheric ammonia. *Appl. Phys. B Lasers Opt.* **2016**, *122*, 1–9. [CrossRef]
64. Maithani, S.; Mandal, S.; Maity, A.; Pal, M.; Pradhan, M. High-resolution spectral analysis of ammonia near 6.2 μm using a cw EC-QCL coupled with cavity ring-down spectroscopy. *Analyst* **2018**, *143*, 2109–2114. [CrossRef]
65. Manne, J.; Lim, A.; Jäger, W.; Tulip, J. Off-axis cavity enhanced spectroscopy based on a pulsed quantum cascade laser for sensitive detection of ammonia and ethylene. *Appl. Opt.* **2010**, *49*, 5302–5308. [CrossRef]
66. Lewicki, R.; Jahjah, M.; Ma, Y.; Stefanski, P.; Tarka, J.; Razeghi, M.; Tittel, F.K. Current status of mid-infrared semiconductor-laser-based sensor technologies for trace-gas sensing applications. *Wonder Nanotechnol. Quantum Optoelectron. Devices Appl.* **2013**, *23*, 597–632. [CrossRef]
67. Lewicki, R.; Kosterev, A.A.; Bakhirkin, Y.A.; Thomazy, D.M.; Doty, J.; Dong, L.; Tittel, F.K.; Risby, T.H.; Solga, S.; Kane, D.; et al. Real Time Ammonia Detection in Exhaled Human Breath with a Quantum Cascade Laser Based Sensor. In *Conference on Lasers and Electro-Optics/International Quantum Electronics Conference*; OSA: Washington, DC, USA, 2009; p. CMS6. [CrossRef]

68. Bielecki, Z.; Stacewicz, T.; Wojtas, J.; Mikołajczyk, J. Application of quantum cascade lasers to trace gas detection. *Bull. Pol. Acad. Sci. Tech. Sci.* **2015**, *63*, 515–525. [CrossRef]
69. Mikołajczyk, J.; Bielecki, Z.; Stacewicz, T.; Smulko, J.; Wojtas, J.; Szabra, D.; Lentka, Ł.; Prokopiuk, A.; Magryta, P. Detection of gaseous compounds with different techniques. *Metrol. Meas. Syst.* **2016**, *23*, 205–224. [CrossRef]
70. Wang, C.; Sahay, P. Breath Analysis Using Laser Spectroscopic Techniques: Breath Biomarkers, Spectral Fingerprints, and Detection Limits. *Sensors* **2009**, *9*, 8230–8262. [CrossRef]
71. Kearney, D.J.; Hubbard, T.; Putnam, D. Breath ammonia measurement in Helicobacter pylori infection. *Dig. Dis. Sci.* **2002**, *47*, 2523–2530. [CrossRef]
72. Smith, D.; Wang, T.; Pysanenko, A.; Španěl, P. A selected ion flow tube mass spectrometry study of ammonia in mouth- and nose-exhaled breath and in the oral cavity. *Rapid Commun. Mass Spectrom.* **2008**, *22*, 783–789. [CrossRef]
73. Buszewski, B.; Grzywinski, D.; Ligor, T.; Stacewicz, T.; Bielecki, Z.; Wojtas, J. Detection of volatile organic compounds as biomarkers in breath analysis by different analytical techniques. *Bioanalysis* **2013**, *5*, 2287–2306. [CrossRef]
74. Thorpe, M.J.; Balslev-Clausen, D.; Kirchner, M.S.; Ye, J. Cavity-enhanced optical frequency comb spectroscopy: Application to human breath analysis. *Opt. Express* **2008**, *16*, 2387. [CrossRef]
75. Stacewicz, T.; Bielecki, Z.; Wojtas, J.; Magryta, P.; Mikołajczyk, J.; Szabra, D. Detection of disease markers in human breath with laser absorption spectroscopy. *Opto-Elektron. Rev.* **2016**, *24*, 29–41. [CrossRef]
76. Assen, A.H.; Yassine, O.; Shekhah, O.; Eddaoudi, M.; Salama, K.N. MOFs for the Sensitive Detection of Ammonia: Deployment of fcu-MOF Thin Films as Effective Chemical Capacitive Sensors. *ACS Sens.* **2017**, *2*, 1294–1301. [CrossRef]
77. Zhang, J.; Ouyang, J.; Ye, Y.; Li, Z.; lin, Q.; Chen, T.; Zhang, Z.; Xiang, S. Mixed-Valence Cobalt(II/III) Metal–Organic Framework for Ammonia Sensing with Naked-Eye Color Switching. *ACS Appl. Mater. Interfaces* **2018**, *10*, 27465–27471. [CrossRef] [PubMed]
78. Galstyan, V.; Bhandari, M.P.; Sberveglieri, V.; Sberveglieri, G.; Comini, E. Metal Oxide Nanostructures in Food Applications: Quality Control and Packaging. *Chemosensors* **2018**, *6*, 16. [CrossRef]
79. Gutowski, P.; Sankowska, I.; Słupiński, T.; Pierścińska, D.; Pierściński, K.; Kuźmicz, A.; Gołaszewska-Malec, K.; Bugajski, M. Optimization of MBE Growth Conditions of In$_{0.52}$Al$_{0.48}$As Waveguide Layers for InGaAs/InAlAs/InP Quantum Cascade Lasers. *Materials* **2019**, *12*, 1621. [CrossRef] [PubMed]
80. Rogalski, A.; Kopytko, M.; Martyniuk, P.; Hu, W. Comparison of performance limits of HOT HgCdTe photodiodes with 2D material infrared photodetectors. *Opto-Electron. Rev.* **2020**, *28*, 82–92. [CrossRef]

© 2020 by the authors. Licensee MDPI, Basel, Switzerland. This article is an open access article distributed under the terms and conditions of the Creative Commons Attribution (CC BY) license (http://creativecommons.org/licenses/by/4.0/).

Review

Trends in Performance Limits of the HOT Infrared Photodetectors

Antoni Rogalski [1], Piotr Martyniuk [1,*], Małgorzata Kopytko [1] and Weida Hu [2]

[1] Faculty of Advanced Technologies and Chemistry, Institute of Applied Physics, Military University of Technology, 2 Kaliskiego St., 00-908 Warsaw, Poland; antoni.rogalski@wat.edu.pl (A.R.); malgorzata.kopytko@wat.edu.pl (M.K.)
[2] State Key Laboratory of Infrared Physics, Shanghai Institute of Technical Physics, Chinese Academy of Sciences, 500 Yu Tian Road, Shanghai 200083, China; wdhu@mail.sitp.ac.cn
* Correspondence: piotr.martyniuk@wat.edu.pl; Tel.: +48-26-183-92-15

Abstract: The cryogenic cooling of infrared (IR) photon detectors optimized for the mid- (MWIR, 3–5 µm) and long wavelength (LWIR, 8–14 µm) range is required to reach high performance. This is a major obstacle for more extensive use of IR technology. Focal plane arrays (FPAs) based on thermal detectors are presently used in staring thermal imagers operating at room temperature. However, their performance is modest; thermal detectors exhibit slow response, and the multispectral detection is difficult to reach. Initial efforts to develop high operating temperature (HOT) photodetectors were focused on HgCdTe photoconductors and photoelectromagnetic detectors. The technological efforts have been lately directed on advanced heterojunction photovoltaic HgCdTe detectors. This paper presents the several approaches to increase the photon-detectors room-temperature performance. Various kinds of materials are considered: HgCdTe, type-II $A^{III}B^V$ superlattices, two-dimensional materials and colloidal quantum dots.

Keywords: HOT IR detectors; HgCdTe; P-i-N; BLIP condition; 2D material photodetectors; colloidal quantum dot photodetectors

1. Introduction

HgCdTe takes the dominant position in infrared (IR) detector technology. This material has triggered the rapid development of the three "detector generations" considered for military and civilian applications and briefly described in the caption of Figure 1. IR detector technology combined with fabrication of epitaxial heterostructure [by molecular beam epitaxy (MBE) and metalorganic chemical vapor deposition (MOCVD)] and photolithographic processes revolutionized the semiconductor industry, thus enabling the design and fabrication of complex focal plane arrays (FPAs). Further their development will relate to implementation of fourth generation staring systems, which the main features are to be: high resolution (pixels > 10^8), multi-band detection, three-dimensional readout integration circuits (3D ROIC), and other integration functions such as polarization/phase sensitivity, better radiation/pixel coupling or avalanche multiplication. The first three generations of imaging systems primarily rely on planar FPAs. Several approaches to circumvent these limitations, including bonding the detectors to flexible or curved molds, have been proposed [1]. Evolution of fourth generation is inspired by the most famous visual systems, which are the biological eyes. Solution based on the Petzval-matched curvature allows the reduction of field curvature aberration. In addition, it combines such advantages as simplified lens system, electronic eye systems and wide field-of-view (FOV) [2,3]. The colloidal quantum dot (CQD) [4] and 2D layered material [5] photodetectors fabricated on flexible substrates exhibit the potential to circumvent technical challenges in the development of the fourth generation IR systems.

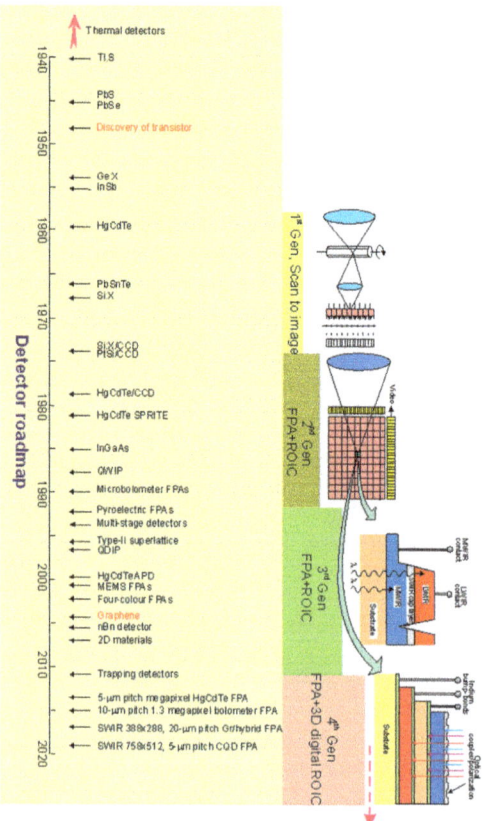

Figure 1. The history of IR detectors and systems development. Four generation systems for military and civilian applications can be considered: first generation (scanning systems), second generation (staring systems—electronically scanned), third generation (staring systems with large number of pixels and two-color functionality), and fourth generation (staring systems with very large number of pixels, multi-color functionality, 3D ROIC, and other on-chip functions; e.g., better radiation/pixel coupling, avalanche multiplication in pixels, polarization/phase sensitivity.

The need for cooling considerably limits more widespread use of IR technology. There are significant attempts to decrease system size, weight, and power consumption (SWaP) to limit IR system's cost and to increase the operating temperature. The invention of microbolometer array was a milestone step in development of IR cameras operating at 300 K. However, microbolometers belong to the class of thermal detectors with limited response time—typically in millisecond range and could not be used in the multiband applications. To omit this limitation, further efforts are directed to increase operating temperature of photon detectors.

Initial efforts in development of the high operating temperatures (HOT) photodetectors were focused on HgCdTe photoconductors and photoelectromagnetic detectors [6]. Many concepts have been implemented and tested to improve the performance of IR photodetectors operating at near 300 K and compiled in References [7–10]. In addition to photoresistors and photodiodes, three other types of IR detectors can operate at near 300 K: magnetic concentration detectors, photoelectromagnetic (or PEM) detectors and Dember effect detectors. The HgCdTe non-equilibrium devices such as the Auger suppressed excluded photoconductors and extracted photodiodes require significant bias what creates excessive 1/f noise.

Up till now, mainly HgCdTe and Sb-based III-V ternary alloys including barrier detectors with type-II superlattices (T2SLs: InAs/GaSb and InAs/InAsSb) have been considered for HOT IR photodetectors. The recently published monograph covers this topic for III-V material systems [11]. In the past decade considerable progress in development of interband quantum cascade infrared photodetectors (IB QCIP) based on T2SLs, 2D material [12] and CQD photodetectors brought their performance close to commercial ones [13].

In 1999 Elliott et al. claimed that there is no fundamental obstacle to reach 300 K operation of photon detectors with background-limited performance even in reduced fields of view [14]. In this paper we attempt to reconsider the performance of different material systems for the HOT detection operation in IR spectral range. Theoretical estimates are collated with experimental data for different photodetectors.

2. Trends in Development of Infrared HOT Photodetectors

As is shown by Piotrowski and Rogalski [15], the IR detectors performance is limited by statistical character of generation-recombination processes in the material. Thermal processes in the device material limit the detectivity, D^* of an optimized IR detector. It can be given by the following equation

$$D^* = k \frac{\lambda}{hc} \left(\frac{\alpha}{G}\right)^{1/2}, \qquad (1)$$

where λ—Wavelength, h—Planck's constant, c—Light speed, α—Absorption coefficient, G—Thermal generation in the active detector's region, k—Coefficient depending on radiation coupling to the detector. α/G is the absorption coefficient to the thermal generation rate ratio and can be considered as the fundamental figure of merit of any material used for IR detectors (α/G ratio could be used to evaluate any material). Among different bulk materials, the narrow gap semiconductors are more suitable for the HOT photodetectors than competitive technologies, such as extrinsic devices, Schottky barrier photodiodes, quantum dot infrared photodetectors (QDIP) and quantum well infrared photodetectors (QWIP) [16]. The high performance of intrinsic photodetectors results from high density of states in the conduction and valence bands (contributing to high IR absorption), and long carrier lifetime (contributing to low thermal generation).

The goal of IR detector technology is the fabrication of HOT photodetector characterized by the dark current lower than the background flux current and $1/f$ noise negligibly lower than to the background flux shot noise [17,18].

In several papers, it was shown that the detector size, d, and F-number (f/#) are the main IR systems parameters [19,20]. Since they depend on $F\lambda/d$ (λ—Wavelength), both influence the detection/identification range, as well as the noise equivalent difference temperature (NEDT) [20]

$$Range = \frac{D\Delta x}{M\lambda}\left(\frac{F\lambda}{d}\right), \qquad (2)$$

$$NEDT \approx \frac{2}{C\lambda(\eta \Phi_B^{2\pi} \tau_{int})}\left(\frac{F\lambda}{d}\right), \qquad (3)$$

where D—Aperture, M—Needed number of pixels to identify a target Δx, C—Scene contrast, η—Quantum efficiency (QE), $\Phi_B^{2\pi}$—Background flux into a 2π FOV, τ_{int}—Integration time. According to the relations (2) and (3), the $F\lambda/d$ parameter could be used for IR system optimization. For the f/1 optics, the smallest practicable detector size should be ~2 µm for the MWIR and ~5 µm for the LWIR, respectively [21]. With more realistic f/1.2 optics, the smallest practicable detector size is ~3 µm and ~6 µm for the MWIR and LWIR, respectively.

Kinch claims that the IR system ultimate cost reduction could only be reached by the 300 K operation of depletion-current limited arrays with pixel densities that are fully consistent with background and diffraction-limited performance due to the system optics [20]. The depletion-current limited P-i-N photodiodes demand long Shockley-Read-Hall (SRH) carrier lifetime, marked as τ_{SRH}, to meet the requirements of a low dark current. The long HgCdTe SRH lifetime makes this material a great candidate for 300 K condition [20].

3. The Ultimate HgCdTe Photodiode Performance

In 2007, the Teledyne Technologies published an empirical expression, called "Rule 07", for estimation of the P-on-n HgCdTe photodiodes dark current versus normalized wavelength-temperature product ($\lambda_c T$) [22]. This metric is closely related to Auger 1 diffusion-limited photodiode with n-type active region doping concentration ~10^{15} cm^{-3}. In the past decade, the Rule 07 has become very popular as a reference level for the other technologies (especially to III-V barrier and T2SLs devices). However, at present stage of technology, the fully depleted background limited HgCdTe photodiodes can reach the level of 300 K dark current considerably lower than predicted by the Rule 07. The discussion below explains exactly this statement.

3.1. SRH Carrier Lifetime

The SRH generation-recombination mechanism determines the carrier lifetimes in both lightly doped n- and p-type HgCdTe in which SRH centres are related to residual impurities and native defects. Kinch et al. in 2005 [23] published that the experimental carrier lifetimes for n-type LWIR HgCdTe range from 2 up to 20 µs at 77 K irrespective of doping <10^{15} cm^{-3}. The MWIR carrier lifetime are substantially longer assuming 2 up to 60 µs. However, several papers published in the last decade have shown τ_{SRH} significantly higher in low temperature range and low doping concentrations, above 200 µs up to even 50 ms versus cut-off wavelength [20] see Table 1. The range of low doping that can be reproducibly obtained in Teledyne Technologies HgCdTe epilayers grown by MBE is ~10^{13} cm^{-3}. Gravrand et al. [24] published that for most tested MWIR photodiodes from CEA Leti and Lynred by Sofradir & Ulis, the estimated SRH carrier lifetimes [from direct measurements (photoconductive or photoluminescence decay) and indirect estimates from current-voltage (I-V) characteristics], are in the range between 10 and 100 µs. Those values are lower than the earlier assessed by US research groups [25]. However, they were estimated for devices with higher doping level in absorber >10^{14} cm^{-3}. However, from just published announcement results, Teledyne Technologies confirmed fabrication of depletion layer limited P-i-N HgCdTe photodiodes with SRH recombination centers exhibiting carrier lifetimes in the range 0.5–10 ms [25].

Table 1. Summary of the SRH carrier lifetimes determined based on current-voltage characteristics (data after reference [20]).

Spectral Range	x Composition	SRH Lifetime (µs)	Temperature (K)
SWIR	0.455	>3000	180
MWIR	0.30	>1000	110
MWIR	0.30	~50,000	89
LWIR	0.225	>100	60

All SRH lifetimes estimated for HgCdTe are usually carried out for temperatures below 300 K. Their extrapolation to 300 K to predict the photodiode operation behavior is questionable. In our estimates we assume τ_{SRH} = 1 ms, which is supported by experimental data reached by Leonardo DRS and Teledyne Technologies research groups.

Figure 2a shows a schematic P-i-N detector energy band profile for a reverse voltage. The active region consists of an undoped i-layer (often called as ν region exhibiting low n-doping) sandwiched between wider bandgap cap (P) and buffer (N) region (see Figure 2b) [26]. Very low doping in the absorber (below 5×10^{13} cm^{-3}) is required to allow full depletion at zero or low reverse voltage. The surrounded wide gap contact layers are designed to reduce the dark current generation from these regions and to prevent tunneling current under reverse bias. Moreover, fully depleted absorber surrounded by wide bandgap regions theoretically reduces 1/f and burst noise. As previously mentioned, the fully depleted P-i-N architecture is compatible with the small pixel size, meeting the requirements of low crosstalk thanks to the built-in electric field [20,26].

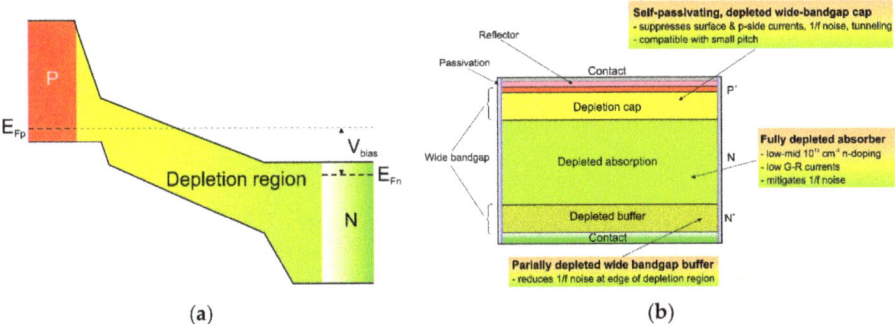

Figure 2. P-i-N detector: (**a**) energy band profile under reverse bias, (**b**) heterojunction architecture (adapted after reference [26]).

In P-i-N design, the choice of absorber thickness should be a trade-off between the response time and QE (or responsivity). To reach short response times, the absorber thickness should be thin and fully depleted. For high QE the absorption region should be thick enough to effectively collect photogenerated carriers. However, to enhance QE while maintaining high response time, an external resonant microcavity was demonstrated [8]. In this design, absorber is placed inside a cavity so that more photons can be absorbed even in low detection volume.

3.2. Dark Current Density

In general, for the fully depleted P-i-N photodiode, the current is built by diffusion from N and P regions (that depends on SRH and Auger generations) and depletion current only determined by SRH generation in the space charge region. Influence of radiative recombination is still debatable but is not considered as a limiting factor of the small pixel HgCdTe photodiodes. Moreover, due to the photon recycling effect, the radiative recombination contribution can be significantly reduced [27]. By that reason in our discussion the radiative recombination is omitted.

The diffusion current of P-i-N HgCdTe photodiode structure arises from the thermal generation of carriers in thick, non-depleted absorber and is dependent on the Auger and SRH generation in n-type semiconductor [20]

$$J_{dif} = \frac{q n_i^2 t_{dif}}{n} \left(\frac{1}{\tau_{A1}} + \frac{1}{\tau_{SRH}} \right), \quad (4)$$

where q—Electron charge, n_i—Intrinsic carrier concentration, n—Electron concentration, t_{dif}—Diffusion region thickness, τ_{A1}—Auger 1 lifetime, and τ_{SRH}—SRH lifetime. Auger 1 lifetime relates to the hole, electron, and intrinsic carrier concentrations, and τ_{A1} is given by equation

$$\tau_{A1} = \frac{2 \tau_{A1}^i n_i^2}{n(n+p)}, \quad (5)$$

where p—Hole concentration and τ_{A1}^i—Intrinsic Auger 1 lifetime. For a low temperature operation or a non-equilibrium active volume, when the majority carrier concentration is held equal to the majority carrier doping level [and intrinsically generated majority carriers are excluded ($p \ll n \approx N_{dop}$)], Equation (5) becomes

$$\tau_{A1} = \frac{2 \tau_{A1}^i n_i^2}{n^2}. \quad (6)$$

The shortest SRH lifetime occurs through centers located approximately at the intrinsic energy level in the semiconductor bandgap. Then, for the field-free region in an n volume ($n \gg p$), τ_{SRH} is given by

$$\tau_{SRH} = \frac{\tau_{no} n_i + \tau_{po}(n + n_i)}{n}, \quad (7)$$

where τ_{no} and τ_{po}—specific SRH lifetimes. At low temperatures, where $n > n_i$, we have $\tau_{SRH} \approx \tau_{po}$. At high temperatures where $n \approx n_i$, we have $\tau_{SRH} \approx \tau_{no} + \tau_{po}$. For a non-equilibrium active volume, $\tau_{SRH} \approx (\tau_{no} + \tau_{po}) n_i / n$ exhibits a temperature dependence given by n_i.

The second component is the depletion current arising from the portion of the absorber that becomes depleted. The depletion current density can be assessed by the relation

$$J_{dep} = \frac{q n_i t_{dep}}{\tau_{no} + \tau_{po}}, \quad (8)$$

where t_{dep}—Depletion region thickness.

The P-i-N HOT detector is characterized by useful properties at reverse voltage. Figure 3 shows the calculated reverse voltage required to completely deplete a 5-µm-thick absorber for selected doping level. For the Rule 7 with doping range about 10^{15} cm^{-3}, a 5-µm-thick absorber can be fully depleted by applying a relatively high reverse bias between 10 V and 30 V. On the other hand, for the doping level reached presently at Teledyne Technologies (~10^{13} cm^{-3}), the 5-µm-thick active layer can be fully depleted for reverse bias from zero up to 0.4 V.

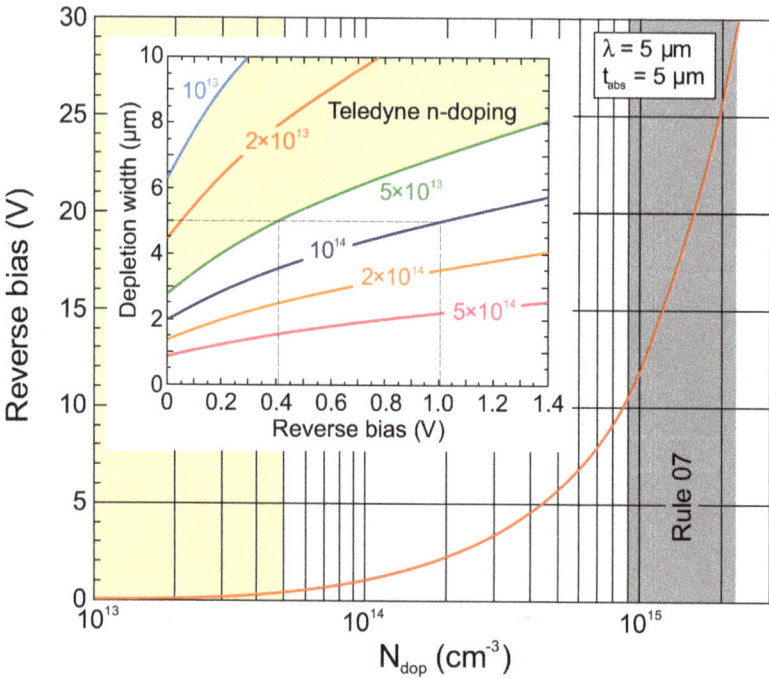

Figure 3. Calculated reverse voltage versus doping concentration required to deplete a 5-µm-thick MWIR HgCdTe absorber. Inset: absorber depletion thickness versus reverse bias and selected doping concentration.

If P-i-N photodiode operates under reverse bias, Auger suppression effect should be considered. This effect is important in HOT condition when $n_i \gg N_{dop}$. In non-equilibrium, large number of intrinsic carriers can be swept-out of the absorber region. It is expected that this impact is larger for lower n-doping levels since n_i is proportionately higher. At very low level of n-type doping (about 10^{13} cm^{-3}) the P-i-N photodiode ultimate performance is influenced by SRH recombination and neither Auger recombination nor Auger suppression.

As is shown in Figure 4, for the sufficiently long SRH carrier lifetime in HgCdTe, the internal photodiode current is limited, and the performance is contributed by the background radiation. Its influence is shown for four background temperatures: 300, 200, 100 and 50 K. Lee et al. suggested to replace Rule 07 by Law 19 corresponding exactly to the background limited curve for room temperature [25]. The internal photodiode current may be several orders of magnitude below Rule 07 versus given cut-off wavelength and operating temperature. It can be also seen that Rule 07 coincides well with theoretically predicted curve for the Auger-suppressed p-on-n photodiode with absorber doping concentration $N_d = 10^{15}$ cm^{-3}.

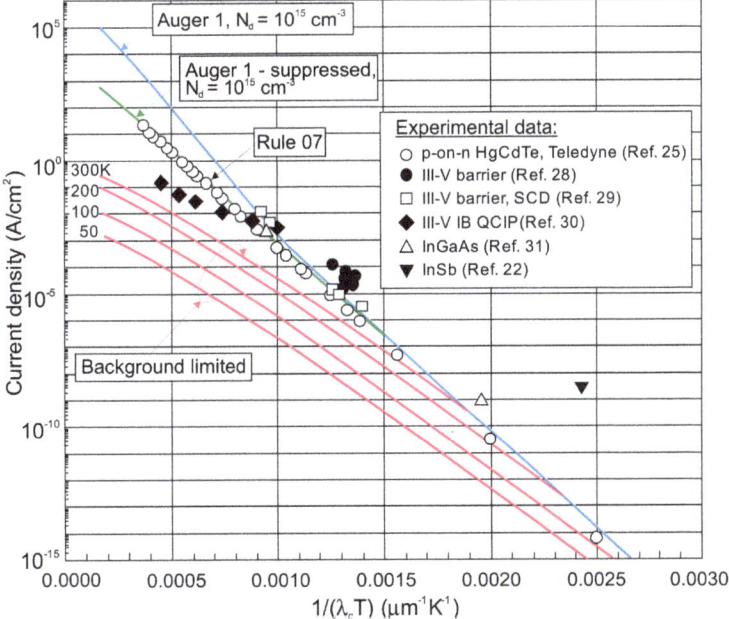

Figure 4. Current density of p-on-n HgCdTe photodiodes versus $1/(\lambda_c T)$ product (adapted after reference [25]). Experimental data is gathered for Teledyne Technologies and alternative technologies [22,25,28–31].

The experimental data for p-on-n HgCdTe photodiodes (Teledyne Technologies) [25] and for III-V barrier detectors (Raytheon Technologies [28] and SCD SemiConductor Devices [29]) operating at about 80 K, and 300 K IB QCIP [30] are presented in Figure 4. It is easy to notice that experimental data for III-V barrier detectors are slightly worse than the p-on-n HgCdTe photodiodes, but III-V IB QCIPs operating at 300 K are even better in LWIR. Figure 4 shows also representative data for both InSb (λ_c = 5.3 μm, T = 78 K) and InGaAs (λ_c = 1.7 and 3.6 μm, T = 300 K) photodiodes. InSb detector is characterized by several orders higher dark current density than HgCdTe one, however for optimal InGaAs detectors the dark current density is close to HgCdTe data [31].

The theoretical simulations presented in Figure 4 indicate that the background limited performance (BLIP) has the most impact on detector's current density for small $1/(\lambda_c T)$ products; in other words for photodiodes operating in LWIR and HOT conditions. HgCdTe photodiodes operating at low temperature become generation-recombination limited due to the SRH centers influence the lifetime in the millisecond range.

Figure 5 shows the current density calculated using Rule 07 (determined for diffusion limited P-on-n photodiodes) and Law 19 (which exactly equals to the background radiation current density) versus temperature for short-wave infrared (SWIR: λ_c = 3 µm), MWIR (5 µm), and LWIR (10 µm) absorber.

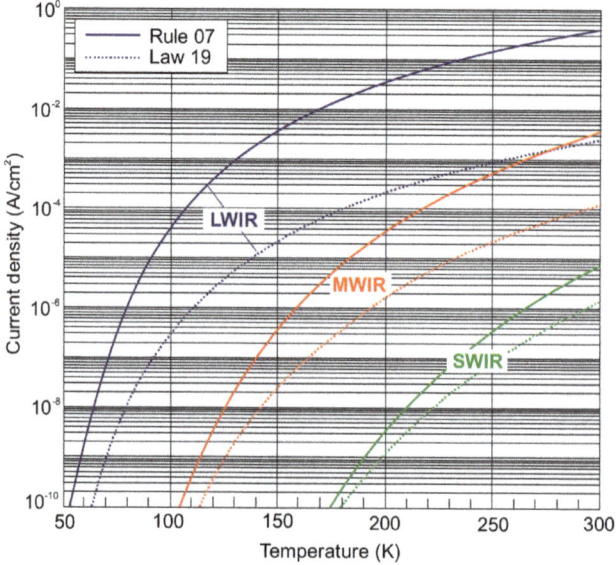

Figure 5. Calculated current density versus temperature using Law 19 and Rule 07 for SWIR (λ_c = 3 µm), MWIR (λ_c = 5 µm), and LWIR (λ_c = 10 µm) HgCdTe absorber.

If the fully depleted P-i-N detector is influenced by the background current, a certain minimal value of SRH lifetime is required. The SRH lifetime calculations were made under condition where depletion dark current equals the background radiation current

$$J_{dep} = J_{BLIP}. \qquad (9)$$

It was assumed that the 5-µm thick absorber is fully depleted.

The SRH lifetime at which the fully depleted P-i-N photodiode reaches the BLIP limit is presented in Figure 6. As shown, the SRH lifetime required to reach BLIP limit decreases versus temperature (nevertheless fully depleted P-i-N photodiodes are particularly interesting in HOT conditions). What more, for LWIR detectors, it is possible to reach BLIP for shorter carrier lifetimes. At 300 K and 5-µm fully depleted thick absorber, these carrier lifetimes are 15 ms for SWIR, 150 µs for MWIR and 28 µs for LWIR, respectively.

The Teledyne Technologies experimentally measured SRH lifetimes for 10-µm cut-off HgCdTe are higher than 100 ms (extracted at 30 K) [26]. Despite the fact that at 300 K the carrier lifetimes are likely to be at least 10 times lower (what results from a high thermal velocity increasing the carrier capture probability by the recombination centre), those low SRH lifetimes enable to reach BLIP limit. This prediction is supported by theoretical simulation presented in reference [32].

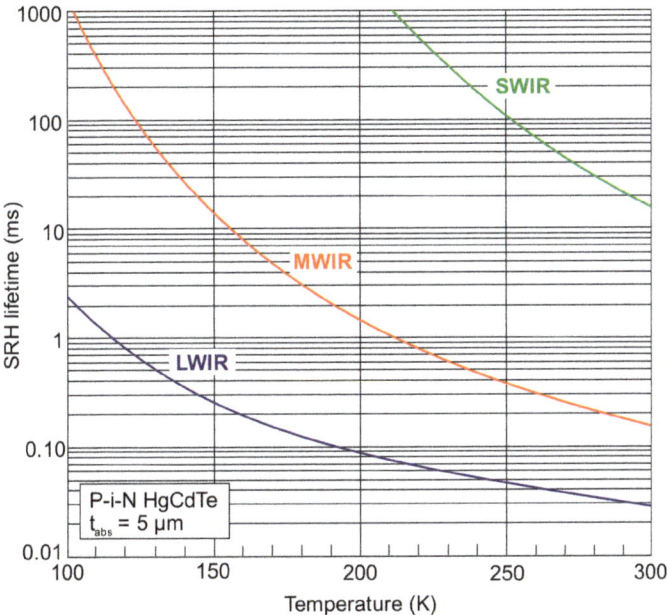

Figure 6. The SRH lifetime versus temperature where fully depleted P-i-N HgCdTe detector depletion dark current equals the background radiation current. The calculations were carried out for SWIR (3 µm), MWIR (5 µm), and LWIR (10 µm) absorbers.

3.3. Detectivity

The detector's detectivity, D^* is related to the current responsivity, R_i [see Equation (13)] and noise current, i_n, and can be given by relation

$$D^* = \frac{R_i}{i_n}. \qquad (10)$$

For the non-equilibrium devices, the i_n value can be calculated assuming thermal Johnson-Nyquist and shot noises contribution by the following expression

$$i_n = \sqrt{\frac{4k_B T}{R_0 A} + 2qJ_{dark}}, \qquad (11)$$

where k_B—Boltzmann constant, $R_0 A$—Dynamic resistance area product and J_{dark}—dark current density.

The performance of P-i-N MWIR HgCdTe photodiode (λ_c = 5 µm) is presented in Figure 7.

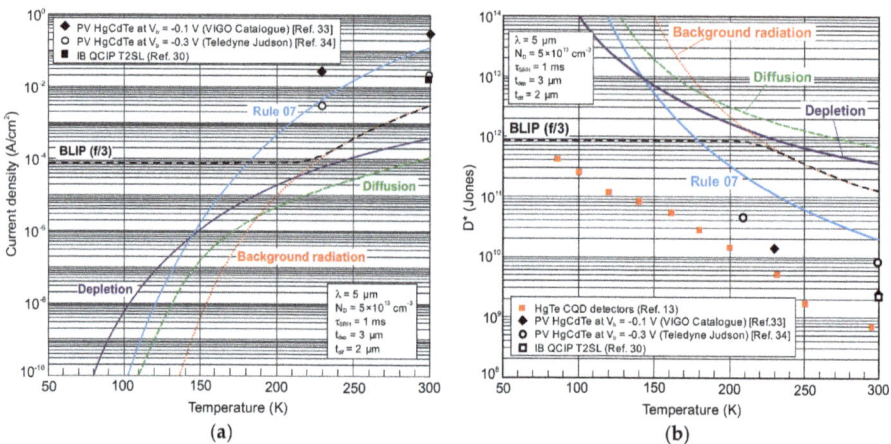

Figure 7. MWIR P-i-N HgCdTe photodiode performance with $\tau_{SRH} = 1$ ms and absorber doping 5×10^{13} cm^{-3}: (**a**) diffusion and depletion current components versus temperature, (**b**) detectivity versus temperature. The thickness of active region: $t = 5$ μm and consists of $t_{dif} = 2$ μm and $t_{dep} = 3$ μm. The experimental data is taken from different sources as marked. PV—Photodiode, CQD—Colloidal quantum dot, IB QCIP—Interband quantum cascade infrared photodetector. Experimental data is taken from [30,33,34].

As is shown in Figure 7a, the Teledyne Judson experimentally measured current densities, at the bias −0.3 V, are close to BLIP (f/3) curve and they are located less than one order of magnitude above this limit [34]. The current density at 300 K is even lower than predicted by Rule 07. The measured current densities presented by VIGO System are close to one order of magnitude higher, however in this case they were measured at lower reverse bias, −0.1 V, with less effective Auger suppression [33]. It is interesting to notice, that the performance of IB QCIP based on T2SLs InAs/GaSb coincides well with upper experimental data for HgCdTe photodiodes at 300 K [30].

Figure 7a shows the diffusion and depletion dark current components versus temperature assuming 1 ms SRH carrier lifetime, 5 μm thick absorber and doping 5×10^{13} cm^{-3}. The diffusion component associated with Auger 1 mechanism is eliminated because of the absence of majority carriers due to exclusion and extraction effects [35,36]. The background radiation calculated assuming f/3 optics has decisive influence on dark current. It should be mentioned here that the background flux current is determined by the net flux through the optics (limited by f/#) plus the flux from the cold shield. This effect is shown by increased BLIP (f/3) influence on dark current at temperature >220 K.

Figure 7b shows calculated detectivity versus temperature for MWIR P-i-N HgCdTe photodiode assuming identical parameters as taken in calculations presented in Figure 7a ($\lambda_c = 5$ μm, $\tau_{SRH} = 1$ ms, $t = 5$ μm, $N_{dop} = 5 \times 10^{13}$ cm^{-3}). The current responsivity was estimated assuming QE = 1 (however typical QE reaches reasonable value ~0.7). As is shown, for MWIR photodiode with 5-μm cut-off wavelength and low doping in active region, detectivity, D^* is limited by background and is about one order of magnitude higher than predicted by the Rule 07. The experimental data given for HgCdTe photodiodes in Teledyne Judson and VIGO System catalogues are more than one order of magnitude below background flux limitation for the f/3 optics.

4. Interband Quantum Cascade Infrared Photodetectors (IB QCIPs)

A low diffusion length, weak absorption and finally low dynamic resistance limit the performance of conventional p-n LWIR HgCdTe HOT detectors with doping concentrations in absorbers > 10^{16} cm^{-3}. The QE is limited since the absorption depth of LWIR ($\lambda > 5$ μm) is much longer than the diffusion length allowing charge carriers photogenerated at distance shorter than the diffusion length to be collected by the contacts. For example, estimates show that for 10.6-µm detector the absorption depth is ~12 µm while the ambipolar diffusion length is less than 2 µm. In consequence, the QE is reduced to ~15% [9].

To overcome above problems, the multiple heterojunction devices based on thin elements connected in series were proposed, where a proper example is a detector with junctions perpendicular to the substrate, introduced in 1995. The multi-heterojunction device shown in Figure 8a contains backside illuminated n$^+$-p-P detectors connected in series. The advantages of such design are a high voltage responsivity, a short response time while on the other hand the response depends on polarization of incident radiation and is nonuniform across the active area.

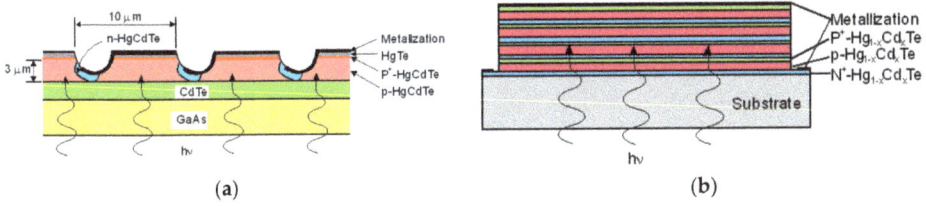

Figure 8. Backside illuminated multiple HgCdTe heterojunction devices: (**a**) junction's planes perpendicular to the surface, and (**b**) 4-cells stacked multiple detector (after reference [8]).

More promising design is the stacked tunnel junctions connected in series as shown in Figure 8b being similar to multi-junction solar cells. Potentially, this device can reach both good QE, high differential resistance, and fast response. As presented, each cell consists of lightly p-type doped absorber and N$^+$/P$^+$ wide-bandgap highly doped contact layers. The heterojunction contacts collect the photogenerated carriers absorbed in every active layer. However, practical problem is related to the resistance of the adjacent N$^+$ and P$^+$ layers.

In the last decade, several designs of the multi-stage IR devices have been developed. They are based on III-V semiconductors and might be now divided into two classes: mentioned earlier interband (IB) ambipolar QCIPs and intersubband (IS) unipolar QCIPs. The first study on IS QCIPs began about two decades ago [11] as the photodetectors were developed from the quantum cascade lasers (QCLs). However, currently IB QCIPs show the higher performance in comparison with IS QCIPs due to the relativity much longer carrier lifetime. The IB QCIPs saturation current density is reported to be almost two orders of magnitude lower than for IS QCIPs [30].

Schematic illustration of IB QCIP is presented in Figure 9 where every single active layer is sandwiched between the relaxation and tunneling layers forming a cascade stage. The thickness of the single active layer should be thinner than the diffusion length to effectively collect all photogenerated carriers. The diffusion length restriction in traditional thick absorber detectors is bypassed by using the discrete absorber design imposing recombination of the photogenerated carriers in the next stage within short transport distance. The single thin absorbers are connected in series and the total thickness of the active layer can be even thicker than the absorption depth. The photocurrent is determined by carriers generated in the single absorber (one stage) and is independent of the number of stages meaning that the photons absorbed in following stages do not increase the net photocurrent but only provide the current continuity through the device. The noises suppression for shorter individual absorbers is the advantage of QCIP design. The QCIP detectivity, D* is influenced by the Johnson and shot noises that is proportional to $\sqrt{N}$ according to the relation [37]

$$i_n = \sqrt{\frac{4k_BT}{NR_0A} + \frac{2qJ_{dark}}{N}},\qquad(12)$$

where N—Number of periods and both dynamic resistance, dark current correspond to one QCIP's period. The optimal number of periods is related to the thickness of the single absorber, d, and the absorption coefficient and could be expressed as $N = (2\alpha d)^{-1}$ in the first order approximation.

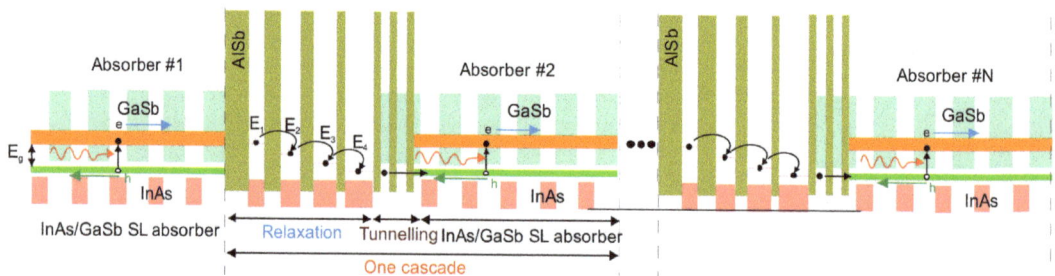

Figure 9. IB QCIP based on T2SL InAs/GaSb active, GaSb/AlSb tunneling and InAs/AlSb relaxation layers (after reference [38]).

The IB QCIPs (with T2SLs absorbers) MBE growth process is challenging where many interfaces and strained thin layers are deposited in structures. Nevertheless, the significant progress has been reached for T2SL based detectors particularly for the LWIR and HOT conditions. They exhibit the capabilities of the IB optical transitions with the exceptional carrier transport properties of the QCIP architectures.

Currently, IB QCIPs has two types of configuration: current-matched (designed to have an equal photocurrent in every single stage) and non-current-matched [39]. Hinkey and Yang described the IB QCIP structure with equal absorbers offering the potential for significant responsivity improvement assuming $\alpha L \leq 0.2$ (αL—Product of the absorption coefficient and the diffusion length) [40]. From a technological point of view, the non-current-matched IB QCIPs (identical absorber thickness in all stages) are simpler to design and grow in comparison to the current-matched ones. The disadvantage of non-current-matched structure is the limited responsivity due to the significant light suppression along the detector's structure. The high electrical gain, lately observed at HOT conditions in these structures, could at least partially compensate in terms of responsivity reduction [41,42].

Despite the development of other technologies, HgCdTe is still the most broadly used adjustable gap semiconductor for IR detectors, to include uncooled operation, and stands as a reference for alternative technologies. Figure 10 demonstrates that T2SL InAs/GaSb IB QCIPs bipolar devices (dashed lines) are proper candidate for HOT conditions. The assessed Johnson-noise limited detectivities under unbiased conditions for IB QCIPs with T2SL InAs/GaSb absorbers (based on the measured R_0A product and responsivity) are comparable with commercially available HgCdTe devices. The performance of both types of detectors is comparable only in SWIR range and IB QCIPs outperform uncooled HgCdTe detectors with a similar LWIR cut-off wavelength. Another advantage of IB QCIPs is related to the III-V semiconductors strong covalent bonding allowing for operation at temperatures close to 400 °C being not possible for HgCdTe.

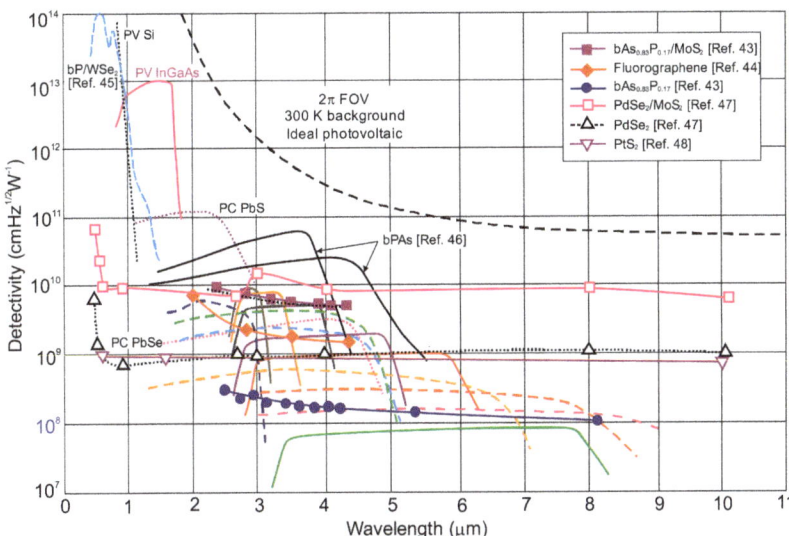

Figure 10. Room-temperature spectral detectivity curves of the commercially available photodetectors [PV Si and InGaAs, PC PbS and PbSe, HgCdTe photodiodes (solid lines reference [33])]. The spectral detectivity curves of new emerging T2SL IB QCIPs are marked by dashed lines (reference 38). Also the experimental data for different types of 2D material photodetectors are included. Experimental data is taken from [43–48]. PC—Photoconductor, PV—Photodiode.

In Figure 10, the representative experimental data for 2D material single photodetectors operating in IR spectral range are also marked. It can be seen that in MWIR the performance of black phosphorous-arsenic (bPAs) photodetectors outperforms commercially available uncooled HgCdTe photodiodes, while in LWIR, the detectivity, D^* of the transition metal dichalcogenide (TMD) photodetectors (PdSe$_2$/MoS$_2$ heterostructure) is the best. More detailed comments about these results are included in Section 5.2.

Figure 11 compares the peak detectivity, D^* for HgCdTe photodiodes [33] and InAs/GaSb T2Sls IB QCIPs [38] operating at 300 K with bPAs and TMD photodetectors. In MWIR the performance of bPAs devices is higher than commercially available uncooled HgCdTe photodiodes, while in LWIR, the detectivity, D^* of the TMD photodetectors is the best. The HgCdTe and IB QCIPs response time at 300 K, typically in the order of nanoseconds, is significantly shorter than for 2D material photodetectors.

Figure 12 gathers the highest detectivity, D^* values published in literature for different types of single element photodetectors operating at room temperature. This fact should be clearly emphasized since detectivity, D^* data marked for commercial photodetectors is typical for pixels of IR FPAs. Figure 12 also presents the fundamental indicator for future trend in development of HOT IR photodetectors. At present stage of HgCdTe technology, the semiempirical Rule 07 is found not to fulfil primary expectations. It is shown that the detectivity, D^* of low-doping P-i-N HgCdTe (5×10^{13} cm^{-3}) photodiodes, operating at 300 K in spectral band above 3 µm, is limited by background radiation (with detectivity, D^* level above 10^{10} Jones, not limited by detector itself) and can be improved more than one order of magnitude in comparison with predicted by Rule 07. Between different material systems used in fabrication of HOT LWIR photodetectors, only HgCdTe can fulfill required expectations: low doping concentration—10^{13} cm^{-3} and high SRH carrier lifetime above 1 ms. In this context, 2D material photodetectors and CQD photodetectors cannot compete with HgCdTe photodiodes. The above assessments provide further inspiration for reaching low-cost and high performance MWIR and LWIR HgCdTe FPAs operating in HOT conditions. The performance of T2SL IB QCIPs is close to HgCdTe photodiodes and quantum cascade photodetectors can operate in temperature > 300 K; however, their disadvantage is a challenging technology and higher fabrication cost.

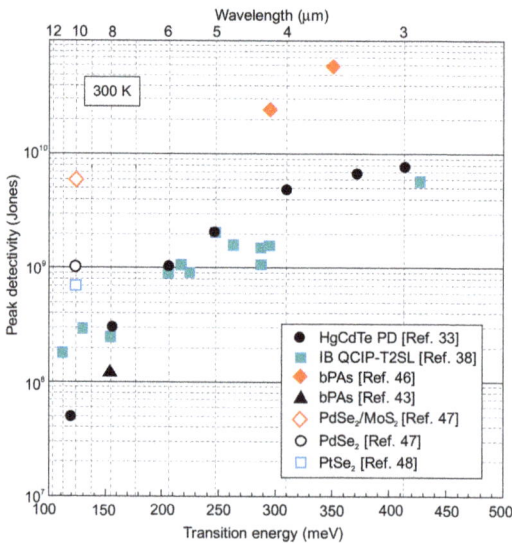

Figure 11. HgCdTe photodiodes, T2SLs InAs/GaSb IB QCIPs and representative 2D material photodetectors peak detectivity, D* comparison for 300 K. The measured data for HgCdTe photodiodes according to the VIGO System catalogue [33]. Data for IB QCIPs extracted from selected papers [38]. Data for selected 2D materials is taken from [33,38,43,46–48].

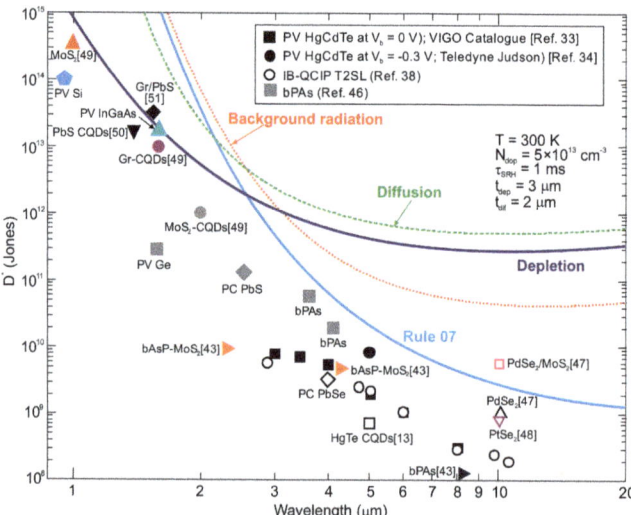

Figure 12. Detectivity, D* versus wavelength for the commercially available room-temperature IR photodetectors (PV Si and Ge, PV InGaAs, PC PbS and PbSe, PV HgCdTe). There is also included experimental data for IB QCIP T2SLs, different type of 2D material and CQD photodetectors taken from literature as marked. The theoretical curves are calculated for the P-i-N HOT HgCdTe devices assuming τ_{SRH} = 1 ms, the absorber doping 5×10^{13} cm^{-3} and the active region thickness t = 5 µm. Experimental data is taken from [13,33,34,38,43,46–51]. PC—Photoconductor, PV—Photodiode.

5. 2D Material Infrared Photodetectors

Graphene and other two-dimensional (2D) materials, due to uncommon electronic and optical properties, are considered to be promising candidates for IR photodetectors [52]. The further development of graphene-based photodetectors is a consequence of the high dark current the gapless material significantly limits the sensitivity. The discovery of new 2D materials with direct energy gaps in a wide spectral range (from the visible to the IR) has set a new direction for detector's design and fabrication. Although the technology readiness, the 2D materials are still at low level of development, the detectors' manufacturability and reproducibility have been challenging (this topic is widely studied in research laboratories around the globe).

Nicolosi et al. [53] distinguished the different types of 2D materials and refined them into different families (see Figure 13) covering a broad range of electrical and optical properties:

- transition metal dichalcogenides (TMDs),
- black phosphorus (bP), metal halides (e.g., PbI_2, $MgBr_2$), metal oxides (such as MnO_2 and MnO_3), double hydroxides, III-Vs (such as InSe and GaS), V-VIs (such as Bi_2Te_3 and Sb_2Se_3), and
- atomically thin hexagonal boron nitride (h-BN, similar to hexagonal sheets of graphene),
- halide perovskites.

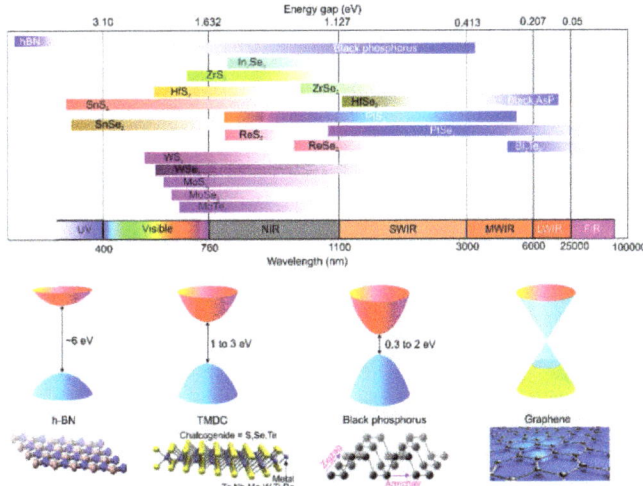

Figure 13. Energy bandgap of the selected layered semiconductors versus wavelength. The energy bandgap exhibits the dependence on the layers number, strain level and chemical doping. FIR—Far infrared; LWIR—Long wavelength infrared; MWIR—Mid wavelength infrared; SWIR—Short wavelength infrared; NIR—Near infrared; UV—Ultraviolet.

2D materials have their roots in layered van der Waals (vdW) solids. Atomic layers are built by in-plane atoms connected by ionic or tight covalent bonds along 2D directions. Each layer is bonded with another by a weak vdW interactions along out-of-plane direction. Such design causes that many of 2D materials could be mechanically exfoliated from bulk single crystals. What more, due to a weak bond between layers, a mixing of different 2D materials together is also possible with the flexibility of the heterostructures.

Energy band profiles of the layered materials differ from their bulk counterparts. Since the material gets thinner from the bulk to the monolayer, e.g., for TMDs the band structure transits from smaller indirect transition to a larger direct due to quantum confinement effects, thus, the bandgap (operating wavelength) can be adjusted by the layers number. Moreover, large strains occurring in these materials strongly affect their optical and electronic properties. A high absorption coefficient of TMDs (typically 10^4–10^6 cm^{-1}) results from dipole transitions between localized d-states and excitonic coupling of such transitions. Thanks to that, >95% of the sunlight is absorbed in sub-micrometer thickness TMD films, while carrier mobility is low (typically less than 250 cm^2/Vs). Despite the fact that the mobility can be improved by increasing the number of TMDs layers, this disadvantage is difficult to circumvent. The carrier density depends on the doping levels and the number of recombination centers, and typically reaches 10^{12} cm^{-2} [54].

In comparison to graphene, TMDs, like molybdenum disulfide (MoS$_2$), tungsten disulfide (WS$_2$) and molybdenum diselenide (MoSe$_2$) is characterized by higher absorption in the visible and NIR ranges. As presented in Figure 13, the 2D bandgap profile is so different what allows to cover a very broad range from the UV to IR. For current status of technology, only graphene-based, black phosphorus-arsenic (bPAs), noble metal dichalcogenides, and bismuthene (like Bi$_2$Te$_3$ and Bi$_2$Se$_3$) are treated as a main player in IR and also THz regions. Since 2D TMDs are limited to UV-NIR, bP can be adjusted to below 0.3 eV by As doping. Due to high mobility, up to 3000 cm^2/Vs, bP is a proper candidate for high sensitivity and fast speed photodetection. Recently published paper indicates that low bandgap 2D noble metal dichalcogenides could be novel platform for 300 K LWIR detectors [47].

From the practical applications point of view, the most important aspect is the stability of the material determining the reliability and lifetime of the device. This is a main disadvantage of most 2D materials. Due to only one or several-atoms thick of the active detector layer, 2D materials are susceptible to ambient environment (especially bP degrades quickly under 300 K conditions). The role of different ambient species has remained debatable [55]. The layered bP devices are still in development stage with many unsolved issues and ideas [56]. On the contrary, the air- stability properties have been demonstrated for noble metal dichalcogenides [47].

5.1. 2D Material Photodetectors: Current Responsivity Versus Response Time

The detector's current responsivity is given by equation

$$R_i = \frac{\lambda \eta}{hc} qg, \tag{13}$$

and is determined by the QE (η) and photoelectrical gain, g. The QE is given by the number of electron-hole pairs generated per incident photon and shows how the detector is coupled to the impinging radiation. The second parameter, the photoelectrical gain describes the number of carriers reaching contacts per one generated pair and shows how well the generated carriers are used to increase photodetector current responsivity. Other symbols of Equation (13) mean: λ—Wavelength, h—Planck constant, q—Electron charge, and c—Light velocity.

In general, the photoelectrical gain is given by

$$g = \frac{\tau}{t_t}, \tag{14}$$

where τ—Carrier lifetime and t_t—Transit time of electrons between the device electrodes. If the drift length, $L_d = v_d \tau$, is less or greater than the distance between electrodes, l, the photoelectrical gain can be less or greater than unity. The value of $L_d > l$ shows that a one carrier swept out by electrode is replaced directly by an equivalent carrier injected by the opposite contact. In this way, a carrier will circulate until it recombines. For the photodiode, the photoelectric gain usually =1, due to separation of minority carriers by the electrical field of depletion region. However, in a hybrid combination of 2D material photodetectors, the photogeneration and carrier transport occur in a separate regions: one for effective light absorption, and the second - to provide fast charge reticulation. In this way, high gain close to 10^8 electrons per one photon, and significant responsivities for SWIR photodetectors have been demonstrated [51].

The simple architecture of hybrid phototransistor, very popular in the 2D material photodetectors design with the fast transfer channel for carriers, is presented in Figure 14. 2D materials with atomic layer thickness are more vulnerable to local electric fields than conventional bulk materials and the photogating effect can strongly modulate the channel conductivity by external gate voltage, V_g. Improvement in the optical gain is particularly important since the QE is suppressed because of the weak absorption in 2D materials. This effect is especially seen in LWIR region, where the light absorption is weak. In the case of hybrid detector shown in Figure 14a, the holes are injected into transporting channel, whereas the electrons remain in the photoactive layer. The injected charges can reticulate even several thousand times before recombination, giving contribution to the gain under illumination. The photocarrier lifetime is enhanced through both the bandgap profile and defect engineering, and at the same time the trapping mechanisms limit the response time of photodetector even to several seconds. The trade-off between improvement in responsivity and response time must be considered during optimization process.

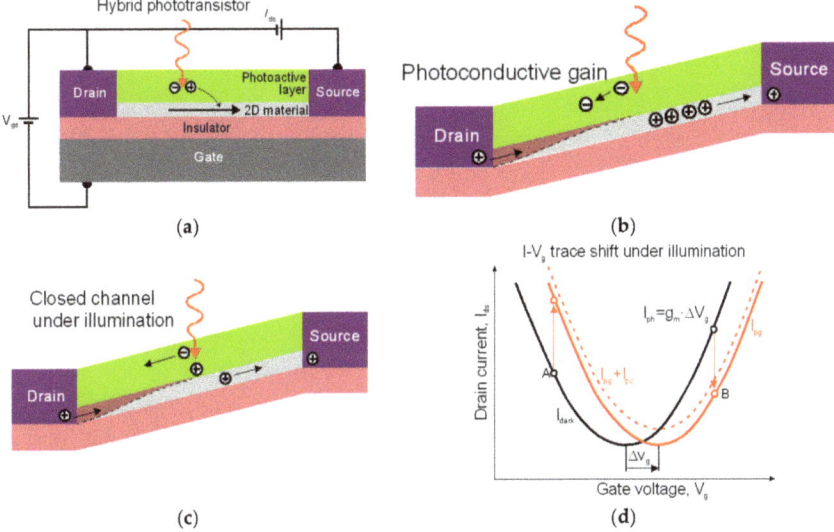

Figure 14. Photogating effect in 2D material photodetectors: (**a**) the operation of hybrid phototransistor, (**b**) closed channel under illumination, (**c**) photoconductive gain, and (**d**) I-V_g trace under illumination.

The photocurrent versus photogating effect can be given by [57]

$$I_{ph} = g_m \Delta V_g, \quad (15)$$

where g_m—Transconductance, ΔV_g—Equivalent photoinduced bias. Figure 14d indicates a shift of the $I_{ds}(V_g)$ trace after the light illumination. Generally, both positive and negative photoconductance effects are observed in hybrid 2D structures and operating points A and B, related to g_m and ΔV_g, perform opposite directions.

Figure 15 compares the graphene-based detectors responsivity operating in visible and NIR with silicon and InGaAs photodiodes commercially available on the market [58,59]. The highest current responsivity, above 10^7 A/W, has been reached for hybrid Gr/quantum dot (QD) photodetectors with enhancement trapped charge lifetimes. As shown, the graphene high mobility along with the extension of the charges lifetime trapped in QDs caused a photodetector responsivity up to seven orders of magnitude higher in relation to the standard bulk photodiodes, where g ≈ 1. Higher responsivity of Si avalanche photodiode (APD), up to 100 A/W, is caused by avalanche process. However, due to the long lifetime of trapped carriers, the response time of 2D material photodetectors is very slow (<10 Hz), what considerably limits real detector functions.

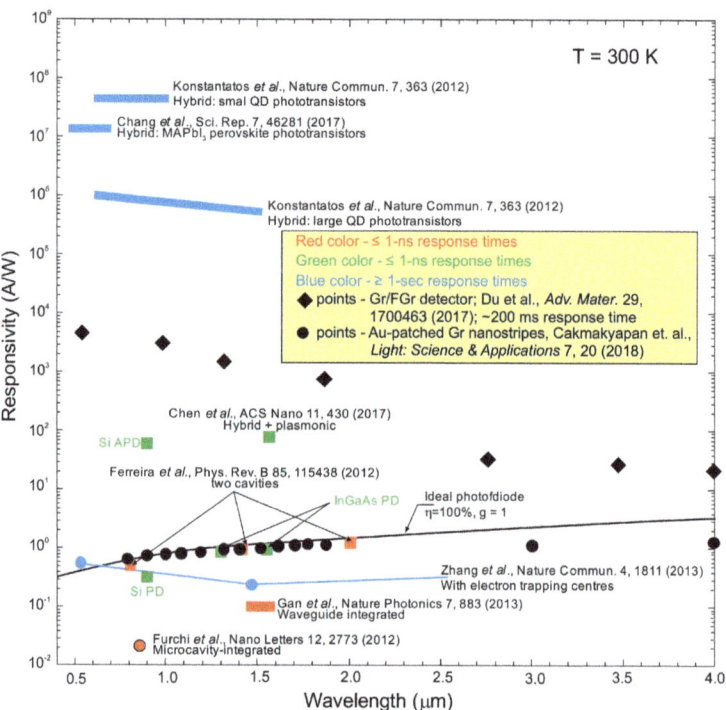

Figure 15. The graphene-based photodetectors spectral responsivity compared with commercial detectors. Black line presents spectral responsivity for ideal photodiode with 100% QE and g = 1. Red and green colors correspond to $\leq$1 ns, while the blue color $\geq$1 second response times. The graphene detectors are labelled with proper reference and brief description. The commercial photodiodes are marked in green (adapted after reference [58,59]).

It is interesting to underline the unusual electrical and optical properties of gold patched graphene nanostripe detectors presented by Cakmakyapan et al. [60]. The photodetector exhibits a spectral response in the ultrabroad range from visible to the IR with high responsivity ranging from 0.6 A/W (for wavelength 800 nm) to 11.65 A/W (for 20 µm) and frequency exceeding 50 GHz. As is shown in Figure 15, its current responsivity (black circles) coincides well with curve (black line) theoretically predicted for ideal photodiode in NIR spectral range.

2D materials show potential for operation in wide spectral range from UV to THz, although majority of them cover visible and SWIR (see Figure 16). Similarly, for graphene photodetectors, both high responsivity and short response time cannot be reached simultaneously in many 2D material-based photodetectors.

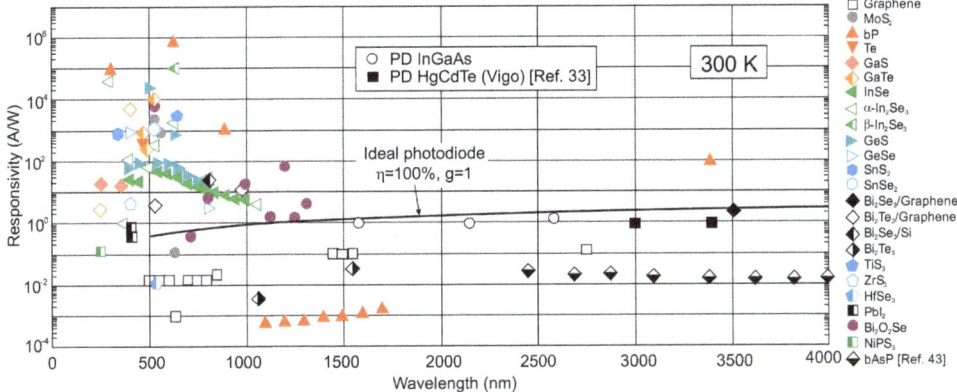

Figure 16. The layered 2D material photodetectors spectral responsivities at 300 K (after reference [43,61]). Black line shows spectral responsivity for ideal photodiode with 100% QE and g = 1. The responsivities of commercially available photodetectors (InGaAs and HgCdTe photodiodes) are presented for comparison reasons.

The two major factors determine the development of the 2D material high sensitivity photodetectors. It is a short carrier lifetime and low absorption in a thin active region (~100–200 nm). In consequence, the broadband operation sets a trade-off between high responsivity and response time. The 2D materials-based detectors display a large variation in their responsivity and response time [62–64] about 9 orders of magnitudes as is shown in Figure 17. In order to improve the IR absorption, the multiple layers instead of the single layer are chosen. In photogating effect photodetectors, 2D materials are used as the fast transfer channel for carriers. However, as is mentioned above, their overall disadvantage is the very slow response time attributed to traps and high capacitance. The response time is typically longer than ~1×10^{-2} ms, what indicates on considerably longer response time in comparison with commercial silicon, InGaAs, and HgCdTe photodiodes, while for HOT LWIR photodiodes is typically tens of nanoseconds. The up-left blank panel on Figure 17 shows the lack of photodetectors with both high responsivity and short response time. Figure 17 summarizes the responsivity and response time of different 2D material photodetectors. It is shown that black phosphorus is a good candidate for fast detection and falls into a region between graphene and TMDs.

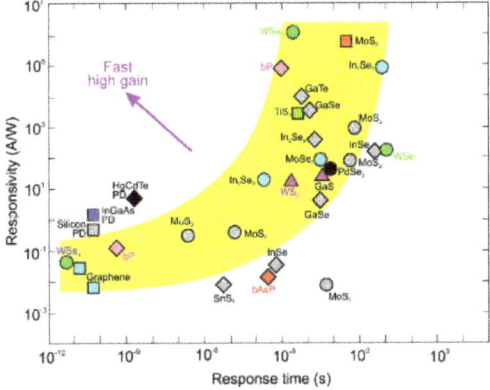

Figure 17. Current responsivity versus response time for HOT 2D material photodetectors in relation to the commercial silicon, InGaAs and HgCdTe detectors (experimental data taken from reference [63]).

5.2. Detectivity: HgCdTe Photodiode Versus 2D Material Photodetectors

Figure 10 presents detectivity, D^* curves gathered from literature for HOT MWIR and LWIR photodetectors both for commercially available devices (PV Si and InGaAs, PC PbS and PbSe, HgCdTe photodiodes) and IB QCIP T2SLs, as well as for 2D material photodetectors. All experimental data gathered in Figure. 10 indicates on sub-BLIP photodetectors performance. As is shown, the literature data for 2D material photodetectors in LWIR above 3 µm is limited to several device structures. The Gr/FGr detector utmost detectivity in MWIR is comparable to HgCdTe. However, especially high detectivity (higher than for HgCdTe photodiodes) is demonstrated for black phosphorus arsenic (bPAs) detectors [46]. Their sensitivity enters the second atmospheric transmission window. Here it must be stressed, that the serious drawback of black phosphorus is surface instability in ambient conditions what can considerably limit its prospective applications [55,65]. More promising is stable TMD photodetectors like $PdSe_2/MoS_2$ heterojunction with record detectivity in LWIR range at room temperature. However, their practical application lies in perfect material synthesis and processing being still under development.

Figure 12 compares the experimental detectivity, D^* published in literature for different types of single element 2D material photodetectors operating at room temperature with theoretically predicted curves for P-i-N HOT HgCdTe detectors. As is presented, the detectivity values for selected 2D material photodetectors are close to data presented for commercial detectors (PV Si and Ge, PV InGaAs, PC PbS and PbSe, PV HgCdTe), and in the case of black phosphorus and TMD detectors are even higher. The enhanced sensitivity of 2D material photodetectors is introduced by bandgap engineering and photogating effect, what degrades the electronic material properties. In consequence, the layered-material photodetectors are characterized by limited linear dynamic range of operation and slow response time.

To summarize the discussion in this section about 2D material IR HOT photodetectors we can conclude that:

- in general, the 2D material IR detectors performance is lower in comparison to the commercially available detectors, especially HgCdTe and new emerging III-V compounds including T2SLs,
- responsivity improvement by using combination of 2D materials with bulks (hybrid photodetectors) owing to the photogating effect causes the limited linear dynamic range due to the charge relaxation time, which lead to decrease in sensitivity with incident optical power,
- responsivity of hybrid and chemically functionalized 2D material photodetectors is comparable with detectors existing on the global market; however, a significant decrease in operating speed (bandwidth) is observed; in general, their response time (millisecond range and longer) is three orders of magnitude longer compared to commercially available photodetectors (microsecond range and shorter) [12],
- the commercialization potential will not just depend on the detector performance, but on the distinct advantages in the ability for fabrication of large scale high quality 2D materials at a low cost,
- experimental data presented by the Teledyne Technologies group [66] support theoretical prediction of background limited P-i-N HgCdTe photodiodes and gives further encouragement for their operation in near room-temperature conditions.

6. Colloidal Quantum Dot Infrared Photodetectors

Research on QD IR photodetectors based on self-assembled epitaxial QDs started in the mid-1990 and were initially very promising. Theoretical estimates carried out by Martyniuk et al. [67] in 2008 indicate that the self-assembled quantum dot infrared photodetectors (QDIPs) are suitable for noncryogenic operation especially in LWIR region. As it happens later, that epitaxial QDs suffer from the size control and low dots density. More recently, an attractive alternative to self-assembled epitaxial QDs has been colloidal quantum dots (CQDs) with better size tunability of optical features and reduction of fabrication cost.

6.1. Brief View

In the last decade, a significant progress in fabrication of CQD photodetectors has been observed. In this approach, an active region is constructed based on 3D quantum confined nanoparticles synthesized by inorganic chemistry. CQDs offer a promising alternative to the single crystal IR materials (InGaAs, InSb, InAsSb, HgCDTe, as well as T2SLs see Figure 18). These nanoparticles could improve CQD photodetectors performance compared to epitaxial QDs due to many aspects gathered in Table 2 [50,68].

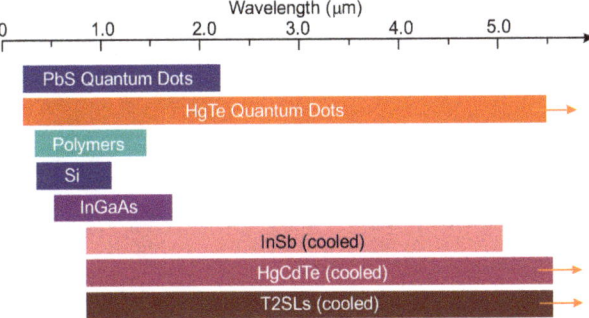

Figure 18. The wavelengths range that can be detected by materials commonly used in imaging applications.

Table 2. CQD photodetectors advantages and disadvantages in comparison with single crystal QD photodetectors.

Advantages	Disadvantages
• control of dot synthesis and absorption spectrum by ability of QD size-filtering, leading to highly-uniform ensembles • much stronger absorption than in Stranski-Krastanov grown QDs due to close-packed of CDs • considerable elimination of strains influencing the growth of epitaxial QDs by better selection of absorber materials • reduction of cost fabrication (using e.g., such solution as spin coating, inject printing, doctor blade or roll-to-roll printing) compared to epitaxial growth • deposition methods are compatible with a variety of flexible substrates and sensing technologies such as CMOS (e.g., direct coating on silicon electronics for imaging)	• inferior chemical stability and electronic passivation of the nanomaterials in comparison with epitaxial materials • bipolar, interband (or excitonic) transitions across the CQD bandgap (e.g., electrons hopping among QDs and holes transport through the polymer) contrary to the intraband transitions in the epitaxial QDs • insulating behaviour due to slow electron transfer through many barrier interfaces in a nanomaterial • problems with long term stability due to the large density of interfaces with atoms presenting different or weaker binding • high level of 1/f noise due to disordered granular systems

CQD photodetectors are typically fabricated using conducting-polymer/nanocrystal blends, or nanocomposites [13,50,69,70]. Nanocomposites often feature narrow-bandgap, II-VI (HgTe, HgSe) [71,72], PbSe or PbS [73–75]. Usually, the reported IR photodetectors use CQDs embedded in conducting polymer matrices, such as poly [2-methoxy-5-(2-ethylhexyloxy]-1,4-phenylenevinylene] (MEH-PPV).

It is expected that the extension of application of CQD-based devices will be significant, especially in IR imaging which is currently dominated by epitaxial semiconductor and hybrid technologies [76]. Hybrid technology, due to the complexity of production stages, reduces yield and increases overall cost. The IR CQD-based photodetectors are an alternative solution without these limitations.

The CQD layers are amorphous what permits fabrication of devices directly onto ROIC substrates, as shown in Figure 19 with no restrictions on pixel or array size and with a day cycle of production. In addition, the monolithic integration of CQD detectors into ROIC does not require any hybridization steps. Individual pixels are defined by the area of the metal pads arranged on the top of ROIC surface. To synthesize colloidal nanocrystals, wet chemistry techniques are used. Reagents are injected into a flask and, the desirable shape and size of nanocrystals are obtained by the control of reagent concentrations, ligand selection, and temperature. This so-called top-surface photodetector offers a 100% fill factor and is compatible with postprocessing at the top of complementary metal-oxide semiconductor (CMOS) electronics.

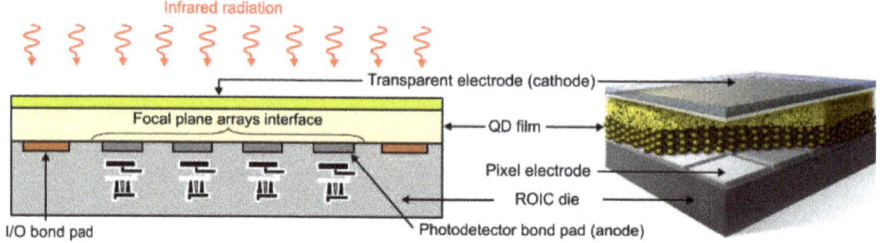

Figure 19. IR monolithic array structure based on CQDs.

The lead chalcogenides CQDs (primarily PbS) are the materials for SWIR photodetectors with detection to 2 µm. The peak can be adjusted using smaller dots by adding NIR bands to hyperspectral visible image sensors or using larger dots to include the InGaAs spectrum of image sensors [75]. From a performance standpoint, SWIR photodetectors based on PbS CQDs have reached detectivity, D^* comparable to commercial InGaAs photodiodes, with a values of >10^{12} Jones at 300 K. HgTe CQDs have opened the MWIR spectral range. Detectivity, D^* between 10^{10} to 10^9 Jones at 5-µm was demonstrated for HgTe CQD devices while maintaining a fast response time at thermoelectric cooling temperatures. It is unlikely that CQD IR detectors will ever reach the performance of currently popular InGaAs, HgCdTe, InSb and T2SL photodiodes.

Recent demonstrations of low-cost SWIR and MWIR CQD imaging arrays have heightened the interest in these devices. For both PbS and HgTe CQD photodetectors integration in camera imaging have been demonstrated [76]. It is expected that the successful implementation of this new class of IR technology may match the broad impact of cheap CMOS cameras that are widely used today. First SWIR cameras built on CQD thin film photodiodes fabricated monolithically on silicon ROICs have been launched [77,78]. The Acuros camera has resolution 1920 × 1080 (2.1 megapixels, 15-µm pixel pitch) and uses 0.4 to 1.7 µm broadband spectral response [77]. The IMEC's prototype imager has resolution of 758 × 512 and 5 µm pixel pitch. The CQD photodiodes on silicon substrate reach an external QE above 60% at 940 nm wavelength, and above 20% at 1450 nm, allowing uncooled operation with dark current comparable to commercial InGaAs photodetectors [78].

At present, CQD cameras are used in newer applications that require high-definition low cost imaging on smaller pixels without extreme sensitivity. It can be predicted that increasing the dot size while maintaining a good mono-dispersion, carrier transport and QE will improve maintaining low noise levels. Due to continuous development of deposition and synthesis techniques, much higher performance will be reached in the future.

6.2. Present Status of CQD Photodiodes

Figure 7b compares the detectivity, D^* temperature dependence versus cut-off wavelength ~5 µm for different material systems including commercially available HgCdTe and

HgTe CQD photodiodes. The gathered experimental data are also included. The estimated detectivity, D* for CQD photodiodes are located below those for HgCdTe photodiodes. As is shown, at temperature above 200 K the theoretically predicted detectivity for HgCdTe photodiodes is limited by background. Rule 07 coincides well with theoretically predicted curve for Auger-suppressed p-on-n HgCdTe photodiode with doping concentration in active region 10^{15} cm^{-3}. As is marked in Section 3.3, at present stage of HgCdTe technology the doping concentration is almost two orders of magnitude lower (mid 10^{13} cm^{-3}).

All experimental data gathered in Figures 12 and 20 indicates on sub-BLIP photodetectors performance. Both figures also clearly show that the detectivity values of CQD photodetectors are inferior in comparison with HgCdTe photodiodes and are generally worse also in comparison with 2D material photodetectors. Moreover, the theoretical predictions indicate on possible further HgCdTe devices performance improvement after decreasing of i-doping level in P-i-N photodiodes. For doping level of 5×10^{13} cm^{-3} the photodiode performance can be limited by background radiation in spectral band above 3 μm. It is shown that in this spectral region, the detectivity, D* is not limited by detector itself, but by background photon noise at a level above 10^{10} Jones in LWIR range (above one order of magnitude above Rule 07).

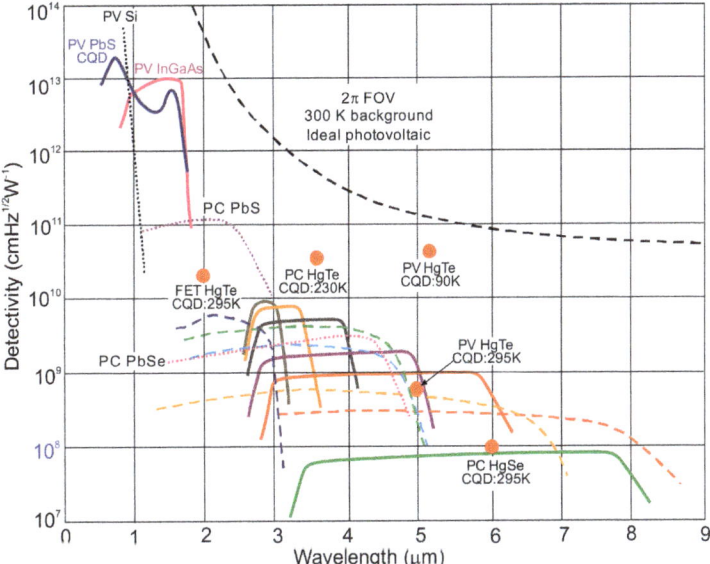

Figure 20. Room-temperature spectral detectivity curves of the commercially available photodetectors [PV Si and InGaAs, PC PbS and PbSe, HgCdTe photodiodes (solid lines reference [33])]. The experimental data for different types of CQD photodetectors are marked by dot points (reference [49,70,78–81]). Also, spectral detectivity of new emerging T2SL IB QCIPs are included [38]. PC—Photoconductor, PV—Photodiode.

7. Conclusions

In the last decade considerable progress in fabrication of SWIR and MWIR 2D material and CQD photodetectors has been demonstrated together with their integration into thermal imaging cameras. At current status of technology, the performance of both types of photodetectors is inferior in comparison with HgCdTe photodiodes. It seems that only PbS CQD photodetectors characterized by multicolor sensitivity and detectivity comparable to InGaAs detectors (which are currently the most common in commercial applications) have been located at the good position in IR material family at present time.

Discovery of graphene in 2004 gave a new impetus on technology development and investigations of 2D layered materials where their uncommon electronic and optical properties make them promising candidates for IR photodetectors. Despite spectacular demonstration of high detectivity like this achieved for black phosphorus layered photodetectors in MWIR spectral range [46] and noble TMD photodetectors like $PdSe_2/MoS_2$ heterojunction with record detectivity in LWIR range at room temperature [47], many challenges remain to be introduced to exploit the distinct advantages of these new materials. The prospect of commercialization of 2D material photodetectors depends on their large-scale integration with existing photonic and electronic platforms like CMOS technologies, high operability, spatial uniformity temporal stability, and affordability. Industry fabrication of devices is in the early stage of development and manufacturability.

In general, pristine narrow gap 2D materials are characterized by weak optical absorption and short carrier lifetime. Various ingenious approaches (electron trap layers, photogating effect with the graphene fast transfer channel) enhance sensitivity, however on the other side, degrade the electronic performance including carrier mobility. In this way high 2D material photodetector sensitivity collides with slow response time what seriously limits their practical applications.

In spite of sixty years development history of HgCdTe, it ultimate HOT performance limit has not been achieved. In order to achieve this goal, the doping concentration below 5×10^{13} cm^{-3} is required. This level of doping concentration has been recently achieved in fully-depleted HgCdTe FPAs by Teledyne Technologies.

At present stage of HgCdTe technology, the semiempirical rule Rule 07 (specified in 2007), widely popular in IR community as a reference for other technologies, was found not to fulfil primary expectations. In this paper, it was shown that the potential properties of HOT HgCdTe photodiodes operating above 3 μm guarantees achieving more than order of magnitude higher detectivity (above 10^{10} Jones) in comparison with value predicted by Rule 07, and this detectivity is limited by background. In this context it is rather difficult to compete 2D material and CQD photodetectors with HgCdTe photodiodes.

Author Contributions: Conceptualization, A.R., P.M.; writing—Original draft preparation, A.R., P.M., M.K. and W.H.; writing—Review and editing, A.R., P.M., M.K. and W.H.; visualization, A.R., P.M., M.K. and W.H.; supervision, A.R. and P.M.; project administration, A.R., P.M. and M.K.; funding acquisition, A.R., P.M. and M.K. All authors have read and agreed to the published version of the manuscript.

Funding: This research was funded by the Polish National Science Centre:, grant number OPUS-UMO-018/31/B/ST7/01541, HARMONIA-UMO-2018/30/M/ST7/00174, OPUS-UMO-2017/27/B/ST7/01507 and UMO-2019/33/B/ST7/00614.

Institutional Review Board Statement: Not applicable.

Informed Consent Statement: Not applicable.

Data Availability Statement: The data presented in this study are available on request from the corresponding author.

Conflicts of Interest: The authors declare no conflict of interest. The funders had no role in the design of the study; in the collection, analyses, or interpretation of data; in the writing of the manuscript, or in the decision to publish the results.

Abbreviations

2D	2 dimensional
3D	3 dimensional
α	The absorption coefficient
APD	Avalanche photodiode
bP	Black phosphorus
bP	Black phosphorus

bPAs	Black phosphorus-arsenic
c	Speed of light
C	Scene contrast
CMOS	Complementary metal-oxide semiconductor
CQD	Colloidal quantum dot
D^*	Detectivity
FIR	Far infrared
FOV	Field-of-view
FPA	Focal plane arrays
g	Photoconductive gain
G	The thermal generation
h	Planck's constant
h-BN	Hexagonal boron nitride
HOT	High operating temperatures
IB QCIP	Interband quantum cascade infrared photodetectors
IR	Infrared radiation
J_{BLIP}	Background radiation current
J_{dark}	Dark current density
J_{dep}	Depletion current
J_{dif}	Diffusion current
λ	Wavelength
LWIR	Long wavelength infrared radiation
MBE	Molecular beam epitaxy
MWIR	Middle wavelength infrared radiation
MOCVD	Metalorganic chemical vapor deposition
NEDT	Noise equivalent difference temperature
n	Electron concentration
n_i	Intrinsic carrier concentration
NIR	Near infrared
p	Hole concentration
PC	Photoconductor
PEM	Photoelectromagnetic
PV	Photodiode
R_i	Current responsivity
q	The electron charge
QE	Quantum efficiency
QD	Quantum dot
QDIP	Quantum dot infrared photodetectors
QWIP	Quantum well infrared photodetectors
R_0A	Dynamic resistance area product
ROIC	Readout integration circuits
SRH	Shockley-Read-Hall
SWaP	Size, weight, and power consumption
SWIR	Short wavelength infrared radiation
T2SLs	Type-II superlattices
THz	Terahertz
TMD	Transition metal dichalcogenide
τ_{A1}	Auger 1 lifetime
τ_{Ai}	Intrinsic Auger 1 lifetime
τ_{int}	Integration time
τ_{po}	Specific SRH lifetimes
τ_{SRH}	SRH lifetime
UV	Ultraviolet
vdW	van der Waals
Φ_B	Background flux

References

1. Iwert, O.; Delabrea, O. The challenge of highly curved monolithic imaging detectors. *Proc. SPIE* **2010**, *7742*, 774227.
2. Jeong, K.-H.; Kim, J.; Lee, L.P. Biologically inspired artificial compound eyes. *Science* **2006**, *312*, 557–561. [CrossRef] [PubMed]
3. Song, Y.M.; Xie, Y.; Malyarchuk, V.; Xiao, J.; Jung, I.; Choi, K.-J.; Liu, Z.; Park, H.; Lu, C.; Kim, R.H.; et al. Digital cameras with designs inspired by the arthropod eye. *Nature* **2013**, *497*, 95–99. [CrossRef] [PubMed]
4. Tang, X.; Ackerman, M.M.; Guyot-Sionnest, P. Colloidal quantum dots based infrared electronic eyes for multispectral imaging. *Proc. SPIE* **2019**, *11088*, 1108803.
5. Lu, Q.; Liu, W.; Wang, X. 2-D Material-Based Photodetectors on Flexible Substrates. In *Inorganic Flexible Optoelectronics: Materials and Applications*; Ma, Z., Liu, D., Eds.; Wiley-VCH Verlag: Weinheim, Germany, 2019; pp. 117–142.
6. Piotrowski, J.; Galus, W.; Grudzień, M. Near room temperature IR photodetectors. *Infrared Phys. Technol.* **1991**, *31*, 1–48. [CrossRef]
7. Piotrowski, J.; Rogalski, A. Photoelectromagnetic, Magnetoconcentration and Dember Infrared Detectors. In *Narrow-Gap II–VI Compounds and Electromagnetic Applications*; Capper, P., Ed.; Chapman & Hall: London, UK, 1997; pp. 506–525.
8. Piotrowski, J.; Rogalski, A. Uncooled long wavelength infrared photon detectors. *Infrared Phys. Technol.* **2004**, *46*, 115–131. [CrossRef]
9. Piotrowski, J.; Rogalski, A. *High-Operating-Temperature Infrared Photodetectors*; SPIE Press: Bellingham, WA, USA, 2007.
10. Piotrowski, J.; Pawluczyk, J.; Piotrowski, A.; Gawron, W.; Romanis, M.; Kłos, K. Uncooled MWIR and LWIR photodetectors in Poland. *Opto-Electron. Rev.* **2010**, *18*, 318–327. [CrossRef]
11. Rogalski, A.; Kopytko, M.; Martyniuk, P. *Antimonide-Based Infrared Detectors. A New Perspective*; SPIE Press: Bellingham, WA, USA, 2018.
12. Rogalski, A.; Kopytko, M.; Martyniuk, P. Two-dimensional infrared and terahertz detectors: Qutlook and status. *Appl. Phys. Rev.* **2019**, *6*, 021316. [CrossRef]
13. Guyot-Sionnest, P.; Ackerman, M.M.; Tang, X. Colloidal quantum dots for infrared detection beyond silicon. *J. Chem. Phys.* **2019**, *151*, 60901. [CrossRef]
14. Elliott, C.T.; Gordon, N.T.; White, A.M. Towards background-limited, room-temperature, infrared photon detectors in the 3–13 μm wavelength range. *Appl. Phys. Lett.* **1999**, *74*, 2881–2883. [CrossRef]
15. Piotrowski, J.; Rogalski, A. Comment on "Temperature limits on infrared detectivities of $InAs/In_xGa_{1-x}Sb$ superlattices and bulk $Hg_{1-x}Cd_xTe$". *J. Appl. Phys.* **1996**, *80*, 2542–2544. [CrossRef]
16. Rogalski, A. *Infrared and Terahertz Detectors*; CRC Press: Boca Raton, FL, USA, 2019.
17. Robinson, J.; Kinch, M.; Marquis, M.; Littlejohn, D.; Jeppson, K. Case for small pixels: System perspective and FPA challenge. *Proc. SPIE* **2014**, *9100*, 91000I.
18. Rogalski, A.; Martyniuk, P.; Kopytko, M. Challenges of small-pixel infrared detectors: A review. *Rep. Prog. Phys.* **2016**, *79*, 046501-1-42. [CrossRef] [PubMed]
19. Holst, G.C.; Lomheim, T.C. *CMOS/CCD Sensors and Camera Systems*; JCD Publishing and SPIE Press: Winter Park, CO, USA, 2007.
20. Kinch, M.A. *State-of-the-Art Infrared Detector Technology*; SPIE Press: Bellingham, WA, USA, 2014.
21. Holst, G.C.; Driggers, R.G. Small detectors in infrared system design. *Opt. Eng.* **2012**, *51*, 96401. [CrossRef]
22. Tennant, W.E.; Lee, D.; Zandian, M.; Piquette, E.; Carmody, M. MBE HgCdTe technology: A very general solution to IR detection, descibrd by 'Rule 07', a very convenient heuristic. *J. Electron. Mater.* **2008**, *37*, 1406–1410. [CrossRef]
23. Kinch, M.S.; Aqariden, F.; Chandra, D.; Liao, P.-K.; Schaake, H.F.; Shih, H.D. Minority carrier lifetime in p-HgCdTe. *J. Electron. Mater.* **2005**, *34*, 880–884. [CrossRef]
24. Gravrand, O.; Rothman, J.; Delacourt, B.; Boulard, F.; Lobre, C.; Ballet, P.H.; Santailler, J.L.; Cervera, C.; Brellier, D.; Pere-Laperne, N.; et al. Shockley-Read-Hall lifetime study and implication in HgCdTe photodiodes for IR detection. *J. Electron. Mater.* **2018**, *47*, 5680–5690. [CrossRef]
25. Lee, D.; Dreiske, P.; Ellsworth, J.; Cottier, R.; Chen, A.; Tallarico, S.; Yulius, A.; Carmody, M.; Piquette, E.; Zandian, M.; et al. Law 19—The ultimate photodiode performance metric. Extended Abstracts. In Proceedings of the 2019 U.S. Workshop on the Physics and Chemistry of II-VI Materials, Chicago, IL, USA, 18–21 November 2019; pp. 13–15.
26. Lee, D.; Carmody, M.; Piquette, E.; Dreiske, P.; Chen, A.; Yulius, A.; Edwall, D.; Bhargava, S.; Zandian, M.; Tennant, W.E. High-operating temperature HgCdTe: A vision for the near future. *J. Electron. Mater.* **2016**, *45*, 4587–4595. [CrossRef]
27. Kopytko, M.; Jóźwikowski, K.; Martyniuk, P.; Rogalski, A. Photon recycling effect in small pixel p-i-n HgCdTe long wavelength infrared photodiodes. *Infrared Phys. Technol.* **2019**, *97*, 38–42. [CrossRef]
28. Rhiger, D.R. Performance comparison of long-wavelength infrared type II superlattice devices with HgCdTe. *J. Elect. Mater.* **2011**, *40*, 1815–1822. [CrossRef]
29. Klipstein, P.C.; Avnon, E.; Azulai, D.; Benny, Y.; Fraenkel, R.; Glozman, A.; Hojman, E.; Klin, O.; Krasovitsky, L.; Langof, L.; et al. Type II superlattice technology for LWIR detectors. *Proc. SPIE* **2016**, *9819*, 98190T.
30. Huang, W.; Rassela, S.M.S.; Li, L.; Massengale, J.A.; Yang, R.Q.; Mishima, T.D.; Santos, M.B. A unified figure of merit for interband and intersubband cascade devices. *Infrared Phys. Technol.* **2019**, *96*, 298–301. [CrossRef]
31. Rogalski, A.; Ciupa, R. Performance limitation of short wavelength infrared InGaAs and HgCdTe photodiodes. *J. Electron. Mater.* **1999**, *28*, 630–636. [CrossRef]
32. Rogalski, A.; Kopytko, M.; Martyniuk, P. Performance prediction of p-i-n HgCdTe long-wavelength infrared HOT photodiodes. *Appl. Optics.* **2018**, *57*, D11–D19. [CrossRef] [PubMed]

33. Available online: https://vigo.com.pl/wp-content/uploads/2017/06/VIGO-Catalogue.pdf (accessed on 30 October 2020).
34. HOT MCT Detectors. Available online: http://www.teledynejudson.com/ (accessed on 30 October 2020).
35. Ashley, T.; Elliott, C.T. Non-equilibrium mode of operation for infrared detection. *Electron. Lett.* **1985**, *21*, 451–452. [CrossRef]
36. Elliott, C.T. Non-equilibrium mode of operation of narrow-gap semiconductor devices. *Semicond. Sci. Technol.* **1990**, *5*, S30–S37. [CrossRef]
37. Gomez, A.; Carras, M.; Nedelcu, A.; Costard, E.; Marcadet, X.; Berger, V. Advantages of quantum cascade detectors. *Proc. SPIE* **2008**, *6900*, 69000J.
38. Rogalski, A.; Martyniuk, P.; Kopytko, M. Type-II superlattice photodetectors versus HgCdTe photodiodes. *Prog. Quantum Electron.* **2019**, *68*, 100228. [CrossRef]
39. Huang, W.; Li, L.; Lei, L.; Massengale, J.A.; Yang, R.Q.; Mishima, T.D.; Santos, M.B. Electrical gain in interband cascade infrared photodetectors. *J. Appl. Phys.* **2018**, *123*, 113104. [CrossRef]
40. Hinkey, R.T.; Yang, R.Q. Theory of multiple-stage interband photovoltaic devices and ultimate performance limit comparison of multiple-stage and single-stage interband infrared detectors. *J. Appl. Phys.* **2013**, *114*, 104506. [CrossRef]
41. Huang, W.; Lei, L.; Li, L.; Massengale, J.A.; Yang, R.Q.; Mishima, T.D.; Santos, M.B. Current-matching versus non-current-matching in long wavelength interband cascade infrared photodetectors. *J. Appl. Phys.* **2017**, *122*, 83102. [CrossRef]
42. Lei, L.; Li, L.; Lotfi, H.; Ye, H.; Yang, R.Q.; Mishima, T.D.; Santos, M.B.; Johnson, M.B. Mid wavelength interband cascade infrared photodetectors with superlattice absorbers and gain. *Opt. Eng.* **2018**, *57*, 11006.
43. Long, M.; Gao, A.; Wang, P.; Xia, H.; Ott, C.; Pan, C.; Fu, Y.; Liu, E.; Chen, X.; Lu, W.; et al. Room temperature high-detectivity mid-infrared photodetectors based on black arsenic phosphorus. *Sci. Adv.* **2017**, *3*, e1700589. [CrossRef] [PubMed]
44. Du, S.; Lu, W.; Ali, A.; Zhao, P.; Shehzad, K.; Guo, H.; Ma, L.; Liu, X.; Pi, X.; Wang, P.; et al. A broadband fluorographene photodetector. *Adv. Mater.* **2017**, *29*, 1700463. [CrossRef]
45. Ye, L.; Wang, P.; Luo, W.; Gong, F.; Liao, L.; Liu, T.; Tong, L.; Zang, J.; Xu, J.; Hu, W. Highly polarization sensitive infrared photodetector based on black phosphorus-on-WSe2 photogate vertical heterostructure. *Nano Energy* **2017**, *37*, 53–60. [CrossRef]
46. Amani, M.; Regan, E.; Bullock, J.; Ahn, G.H.; Javey, A. Mid-wave infrared photoconductors based on black phosphorus-arsenic alloys. *ACS Nano* **2017**, *11*, 11724–11731. [CrossRef]
47. Long, M.; Wang, Y.; Wang, P.; Zhou, X.; Xia, H.; Luo, C.; Huang, S.; Zhang, G.; Yan, H.; Fan, Z.; et al. Palladium diselenide long-wavelength infrared photodetector with high sensitivity and stability. *ACS Nano* **2019**, *13*, 2511–2519. [CrossRef]
48. Yu, X.; Yu, P.; Wu, D.; Singh, B.; Zeng, Q.; Lin, H.; Zhou, W.; Lin, J.; Suenaga, K.; Liu, Z.; et al. Atomically thin noble metal dichalcogenide: A broadband mid-infrared semiconductor. *Nat. Commun.* **2018**, *9*, 1545. [CrossRef]
49. Konstantatos, G. Current status and technological prospect of photodetectors based on two-dimensional materials. *Nat. Commun.* **2018**, *9*, 5266. [CrossRef]
50. Konstantatos, G.; Sargent, E.H. Solution-processed quantum dot photodetectors. *Proc. IEEE.* **2009**, *97*, 1666–1683. [CrossRef]
51. Konstantatos, G.; Badioli, M.; Gaudreau, L.; Osmond, J.; Bernechea, M.; Garcia de Arquer, F.P.; Gatti, F.; Koppens, F.H.L. Hybrid graphene-quantum dot phototransistors with ultrahigh gain. *Nat. Nanotechnol.* **2012**, *7*, 363–368. [CrossRef] [PubMed]
52. Morgan, H.; Rout, C.S.; Late, D.J. (Eds.) *Fundamentals and Sensing Applications of 2D Materials*; Woodhead Publishing Series in Electronic and Optical Materials; United Kingdom Elsevier: London, UK, 2019.
53. Nicolosi, V.; Chhowalla, M.; Kanatzidis, M.G.; Strano, M.S.; Coleman, J.N. Liquid exfoliation of layered materials. *Science* **2013**, *340*, 1226419. [CrossRef]
54. Yang, Z.; Dou, J.; Wang, M. Graphene, transition metal dichalcogenides, and perovskite photodetectors. In *Two-Dimensional Materials for Photodetector*; IntechOpen: Rijeka, Croatia, 2018. [CrossRef]
55. Wang, X.; Sun, Y.; Liu, K. Chemical and structural stability of 2D layered materials. *2D Mater.* **2019**, *6*, 42001. [CrossRef]
56. Ling, X.; Wang, H.; Huang, S.; Xia, F.; Dresselhaus, M.S. The renaissance of black phosphorus. *Proc. Natl. Acad. Sci. USA* **2015**, *112*, 4523–4530. [CrossRef] [PubMed]
57. Wang, P.; Xia, H.; Li, Q.; Wang, F.; Zhang, L.; Li, T.; Martyniuk, P.; Rogalski, A.; Hu, W. Sensing infrared photons at room temperature: From bulk materials to atomic layers. *Small* **2019**, *46*, 1904396. [CrossRef]
58. Currie, M. *Applications of Graphene to Photonics*; Report NRL/MR/5650-14-9550; Naval Research Laboratory: Washington, DC, USA, 2014.
59. Rogalski, A.; Kopytko, M.; Martyniuk, P. 2D material infrared and terahertz detectors: status and outlook. *Opto-Electron. Rev.* **2020**, *28*, 107–154.
60. Cakmakyapan, S.; Lu, P.K.; Navabi, A.; Jarrahi, M. Gold-patched graphene nano-stripes for high-responsivity and ultrafast photodetection from the visible to infrared regime. *Light Sci. Appl.* **2018**, *7*, 20. [CrossRef]
61. Wang, F.; Wang, Z.; Yin, L.; Cheng, R.; Wang, J.; Wen, Y.; Shifa, T.A.; Wang, F.; Zhang, Y.; Zhan, X.; et al. 2D library beyond graphene and transition metal dichalcogenides: A focus on photodetection. *Chem. Soc. Rev.* **2018**, *47*, 6296–6341. [CrossRef]
62. Buscema, M.; Island, J.O.; Groenendijk, D.J.; Blanter, S.I.; Steele, G.A.; Van der Zant, H.S.J.; Castellanos-Gomez, A. Photocurrent generation with two-dimensional van der Waals semiconductor. *Chem. Soc. Rev.* **2015**, *44*, 3691–3718. [CrossRef]
63. Wang, J.; Fang, H.; Wang, X.; Chen, X.; Lu, W.; Hu, W. Recent progress on localized field enhanced two-dimensional material photodetectors from ultraviolet-visible to infrared. *Small* **2017**, *13*, 1700894. [CrossRef]
64. Long, M.; Wang, P.; Fang, H.; Hu, W. Progress, challenges, and opportunities for 2D material based photodetectors. *Adv. Funct. Mater.* **2018**, *29*, 1803807. [CrossRef]

65. Xu, Y.; Shi, Z.; Shi, X.; Zhang, K.; Zhang, H. Recent progress in black phosphorus and black-phosphorus-analogue materials: Properties, synthesis and applications. *Nanoscale* **2019**, *11*, 14491–14527. [CrossRef] [PubMed]
66. Lee, D.; Dreiske, P.; Ellsworth, J.; Cottier, R.; Chen, A.; Tallarico, S.; Barr, H.; Tcheou, H.; Yulius, A.; Carmody, M.; et al. Performance of MWIR and LWIR fully-depleted HgCdTe FPAs. Extended Abstracts. In Proceedings of the 2019 U.S. Workshop on the Physics and Chemistry of II-VI Materials, Chicago, IL, USA, 18–21 November 2019; pp. 189–190.
67. Martyniuk, P.; Rogalski, A. Comparison of performance of quantum dot and other types infrared photodetectors. *Proc. SPIE* **2008**, *6940*, 694004.
68. Stiff-Roberts, A.D. Quantum-dot infrared photodetectors: A review. *J. Nanophotonics* **2009**, *3*, 31607. [CrossRef]
69. Ginger, D.S.; Greenham, N.C. Photoinduced electron transfer from conjugated polymers to CdSe nanocrystals. *Phys. Rev. B* **1999**, *59*, 10622–10629. [CrossRef]
70. Garcia de Arquer, F.P.; Armin, A.; Meredith, P.; Sargent, E.H. Solution-processed semiconductors for next-generation photodetectors. *Nat. Rev. Mater.* **2017**, *2*, 16100. [CrossRef]
71. Guyot-Sionnest, P.; Roberts, J.A. Background limited mid-infrared photodetection with photovoltaic HgTe colloidal quantum dots. *Appl. Phys. Lett.* **2015**, *107*, 91115. [CrossRef]
72. Buurma, C.; Ciani, A.J.; Pimpinella, R.E.; Feldman, J.S.; Grein, C.H.; Guyot-Sionnes, P. Advances in HgTe colloidal quantum dots for infrared detectors. *J. Electron. Mater.* **2017**, *46*, 6685–6688. [CrossRef]
73. De Iacovo, A.; Venettacci, C.; Colace, L.; Scopa, L.; Foglia, S. PbS colloidal quantum dot photodetectors operating in the near infrared. *Sci. Rep.* **2016**, *6*, 37913. [CrossRef]
74. Thambidurai, M.; Jjang, Y.; Shapiro, A.; Yuan, G.; Xiaonan, H.; Xuechao, Y.; Wang, G.J.; Lifshitz, E.; Demir, H.V.; Dang, C. High performance infrared photodetectors up to 2.8 μm wavelength based on lead selenide colloidal quantum dots. *Opt. Mater. Express* **2017**, *7*, 2336. [CrossRef]
75. Malinowski, P.E.; Georgitzikis, E.; Maes, J.; Vamvaka, I.; Frazzica, F.; Van Olmen, J.; De Moor, P.; Heremans, P.; Hens, Z.; Cheyns, D. Thin-film quantum dot photodiode for monolithic infrared image sensors. *Sensors* **2017**, *17*, 2867. [CrossRef] [PubMed]
76. Hafiz, S.B.; Scimeca, M.; Sahu, A.; Ko, D.-K. Colloidal quantum dots for thermal infrared sensing and imaging. *Nano Converg.* **2019**, *6*, 7. [CrossRef] [PubMed]
77. Available online: https://ibv.vdma.org/documents/256550/27019077/2018-11-07_Stage1_1030_SWIR+Vision+Systems.pdf/ (accessed on 30 October 2020).
78. Available online: https://optics.org/news/10/10/38 (accessed on 30 October 2020).
79. Lhuillier, E.; Guyot-Sionnest, P. Recent progress in mid infrared nanocrystal optoelectronics. *IEEE J. Sel. Top. Quantum Electron.* **2017**, *23*, 6000208. [CrossRef]
80. Chen, M.; Lu, H.; Abdelazim, N.M.; Zhu, Y.; Wang, Z.; Ren, W.; Kershaw, S.V.; Rogach, A.L.; Zhao, N. Mercury telluride quantum dot based phototransistor enabling high-sensitivity room-temperature photodetection at 2000 nm. *ACS Nano* **2017**, *11*, 5614–5622. [CrossRef] [PubMed]
81. Livache, C.; Martinez, B.; Goubet, N.; Ramade, J.; Lhuillier, E. Road map for nanocrystal based infrared photodetectors. *Front. Chem.* **2018**, *6*, 575. [CrossRef] [PubMed]

Review

Low-Light Photodetectors for Fluorescence Microscopy

Hiroaki Yokota [1,*], Atsuhito Fukasawa [2], Minako Hirano [1] and Toru Ide [3]

1. Biophotonics Laboratory, The Graduate School for the Creation of New Photonics Industries, 1955-1 Kurematsu-cho, Nishi-ku, Hamamatsu-shi, Shizuoka 431-1202, Japan; hirano37@gpi.ac.jp
2. Electron Tube Division, Hamamatsu Photonics K.K., 314-5 Shimokanzo, Iwata-shi, Shizuoka 438-0193, Japan; fukasawa@etd.hpk.co.jp
3. Graduate School of Interdisciplinary Science and Engineering in Health Systems, Okayama University, 3-1-1 Tsushima-naka, Kita-ku, Okayama-shi, Okayama 700-8530, Japan; ide@okayama-u.ac.jp
* Correspondence: yokota@gpi.ac.jp

Abstract: Over the years, fluorescence microscopy has evolved and has become a necessary element of life science studies. Microscopy has elucidated biological processes in live cells and organisms, and also enabled tracking of biomolecules in real time. Development of highly sensitive photodetectors and light sources, in addition to the evolution of various illumination methods and fluorophores, has helped microscopy acquire single-molecule fluorescence sensitivity, enabling single-molecule fluorescence imaging and detection. Low-light photodetectors used in microscopy are classified into two categories: point photodetectors and wide-field photodetectors. Although point photodetectors, notably photomultiplier tubes (PMTs), have been commonly used in laser scanning microscopy (LSM) with a confocal illumination setup, wide-field photodetectors, such as electron-multiplying charge-coupled devices (EMCCDs) and scientific complementary metal-oxide-semiconductor (sCMOS) cameras have been used in fluorescence imaging. This review focuses on the former low-light point photodetectors and presents their fluorescence microscopy applications and recent progress. These photodetectors include conventional PMTs, single photon avalanche diodes (SPADs), hybrid photodetectors (HPDs), in addition to newly emerging photodetectors, such as silicon photomultipliers (SiPMs) (also known as multi-pixel photon counters (MPPCs)) and superconducting nanowire single photon detectors (SSPDs). In particular, this review shows distinctive features of HPD and application of HPD to wide-field single-molecule fluorescence detection.

Keywords: low-light photodetectors; fluorescence microscopy; time-resolved fluorescence microscopy; hybrid photodetector (HPD); single-molecule fluorescence detection

1. Introduction

The development of the light microscope has enabled investigation of the fine structures of biological specimens under magnification. In recent decades, fluorescence microscopy—a form of highly sensitive optical microscopy—has evolved and is a necessary element of life science studies [1–3]. Microscopy has elucidated biological processes in vitro and in vivo, and also enabled tracking of biomolecules in real time. Development of highly sensitive photodetectors and light sources, in addition to evolution of various illumination methods and fluorophores, has helped microscopy acquire single-molecule fluorescence sensitivity, enabling single-molecule fluorescence imaging and detection [4–7]. Single-molecule fluorescence microscopy led to the emergence of super-resolution microscopy [8–10].

Low-light photodetectors used in fluorescence microscopy are classified into two categories: point photodetectors and wide-field photodetectors [5,11]. Point photodetectors, notably photomultiplier tubes (PMTs), are most commonly used in laser scanning microscopy (LSM). The detectors are also used in point-like excitation and detection to study freely diffusing biomolecules, such as protein molecules and nucleic acids (DNA and RNA) in solution [11]. Wide-field photodetectors, such as electron-multiplying charge-coupled devices (EMCCDs) and scientific complementary metal-oxide-semiconductor (sCMOS) cameras, are used in wide-field illumination and detection to study surface-immobilized or slowly-diffusing biomolecules and organelles. These biomolecules and organelles include protein molecules, nucleic acids, and lipids, and nucleus and mitochondria, respectively. The two photodetectors are distinct in many aspects. Point photodetectors have high temporal resolution (high sampling frequency) but have no spatial resolution without scanning. Wide-field photodetectors, in contrast, have spatial resolution (typically sub-micrometer precision) but are limited to relatively low frame rates. This review focuses on low-light point photodetectors and presents their fluorescence microscopy applications and recent progress.

In fluorescence microscopy, the fluorescence emission can be characterized not only by intensity and position but also by lifetime [12]. Fluorescence microscopy uses two fluorescence detection methods: steady-state and time-resolved fluorescence detection. Time-resolved measurements contain more information than is available from steady-state measurements. Fluorescence lifetime measurement by time-resolved detection provides data that is independent of fluorophore concentration and allows us to obtain information on the local ambient environment around the fluorophore, such as pH, ion concentrations, temperature, and fluorescence resonance energy transfer (FRET) efficiency [13]. The laser scanning fluorescence microscope, an indispensable imaging device in the biological sciences, is one of the most widely used fluorescence microscopy. LSM with a confocal illumination setup provides a base for various fluorescence microscopes using steady-state and time-resolved fluorescence detection, such as two-photon microscopy and fluorescence lifetime imaging microscopy (FLIM). This review introduces these point detectors, describes their operating principles, and compares their specifications. These photodetectors include conventional PMTs, single photon avalanche diodes (SPADs), hybrid photodetectors (HPDs), in addition to newly emerging multi-pixel photon counters (MPPCs) (also known as silicon photomultipliers (SiPMs)) and superconducting nanowire single photon detectors (SSPDs). In particular, this review shows distinctive features of HPD, and notes the applications of HPD to wide-field single-molecule fluorescence detection and the development of multi-pixel photodetectors.

2. Fluorescence Microscopy

Fluorescence measurements are characterized by their high sensitivity, up to single-molecule detection. Because biological samples commonly exhibit low contrast, fluorescence microscopy makes good use of fluorescence phenomenon to enhance the contrast. Fluorescence microscopy acquires data of target biological samples through fluorescence emissions characterized not only by intensity and position, but also by lifetime. Fluorescence microscopy uses the two fluorescence detection methods: steady-state and time-resolved fluorescence detection. LSM with a confocal illumination setup provides a base for various fluorescence microscopes using steady-state and time-resolved fluorescence detection, such as two-photon microscopy and FLIM.

2.1. Fluorescence

Fluorescence is a photophysical phenomenon of the emission of light through the excitation of a fluorophore from the ground state to an excited electronic state upon the absorption of light, with the light energy equivalent to the energy gap between the two states [10,12]. Figure 1 shows a Jablonski diagram that illustrates the electronic energy levels of a fluorophore and the transitions between them are represented by arrows. S_0, S_1, and S_2 represent the singlet ground, and first and second excited electronic states, respectively. The vibrational ground states and higher vibrational states of each electronic state are illustrated with black and gray lines, respectively. The transition from the ground state to the excited state by the light absorption occurs in less than 10^{-15} s. A fluorophore is usually excited to some higher vibrational level of the excited state. The electron usually rapidly relaxes to the lowest vibrational level of S_1. This process is called internal conversion and generally occurs within 10^{-12} s. Then, the excited state is relaxed to the ground state in a few nanoseconds (10^{-9} s), which is accompanied by the radiation of the fluorescence emission. Because internal conversion is generally complete prior to emission, the last relaxation step to the ground state accounts for most of the overall process. Thus, the relaxation time is called the fluorescence lifetime. The wavelength of the fluorescence is longer than the excitation wavelength because the energy from the absorbed photon is partially lost via non-radiative decay. This shift in wavelength is called the Stokes shift.

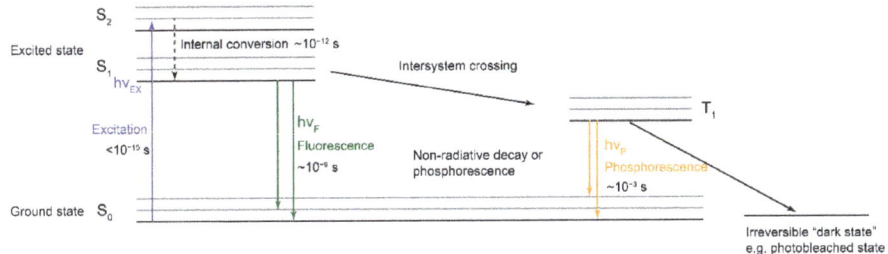

Figure 1. Jablonski diagram for the electronic energy levels of a fluorophore and the transitions between them.

An electron in the S_1 state can also flip its spin thus creating the first triplet state T_1, which is termed intersystem crossing. Transition from T_1 to S_0 is forbidden, thus the rate constants for triplet emission (phosphorescence) are several orders of magnitude smaller than those for fluorescence, which results in a long-lived dark state called blinking. While in the T_1 state, the fluorophore may experience photobleaching, which is an irreversible fluorescence switching-off process.

Table 1 shows characteristics of several fluorophores commonly used in fluorescence microscopy, including dyes, a quantum dot (Qdot) [14–16], and a fluorescent protein [17,18]. Photobleaching can be a representative photostability indicator of a fluorophore and the higher resistance to photobleaching allows longer or brighter fluorescence observation.

Table 1. Characteristics of fluorophores commonly used in fluorescence microscopy.

Fluorophore	Typical Excitation Wavelength (nm)	Emission Wavelength (nm)	Molecular Extinction Coefficient ($M^{-1}cm^{-1}$)	Quantum Yield	Fluorescence Lifetime (ns)	Resistance to Photoblea-Ching
Cy3	532	570	150,000 (550 nm)	0.15	0.2	++
Cy5	633	670	250,000 (649 nm)	0.28	0.9	++
Qdot655	532, 633	~655	2,100,000 (532 nm)	0.44	~20	+++
Green fluorescent protein (GFP)	488	507	56,000 (484 nm)	0.60	2.3	+

The markings +, ++, and +++ indicate "poor," "moderate," and "excellent," respectively. Adapted with permission from [7].

2.2. Laser Scanning Fluorescence Microscopy (LSM)

Laser scanning fluorescence microscopy (LSM) is one of the most widely used forms of biological fluorescence microscopy. LSM creates fluorescence images by sequentially recording the fluorescence intensity of each pixel by scanning a focused laser beam across a specimen using confocal optics to obtain only in-focus fluorescence. LSM with a confocal illumination setup provides a base for various fluorescence microscopes using steady-state and time-resolved fluorescence microscopy. Figure 2 shows schematics of confocal microscopy and fluorescence microscopy with confocal optics.

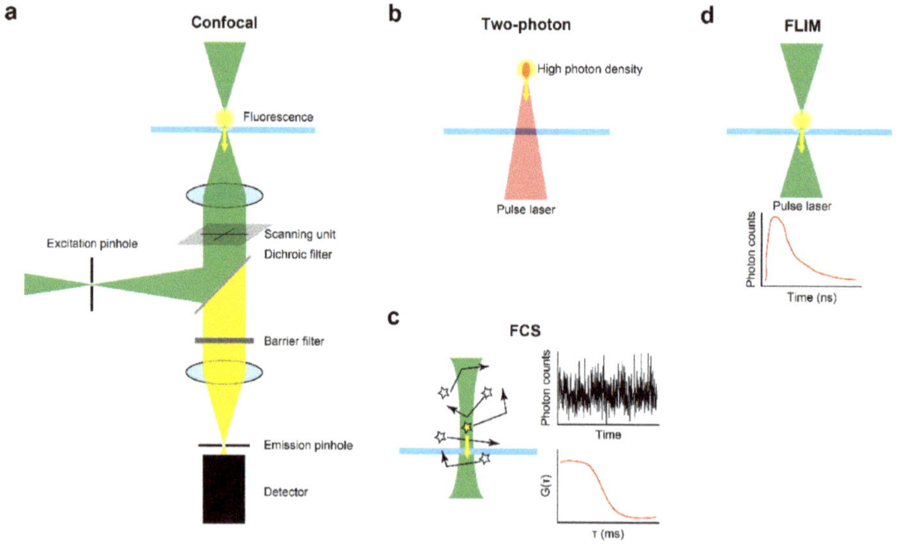

Figure 2. Schematics of confocal microscopy and fluorescence microscopy with confocal optics: (**a**) confocal microscopy; (**b**) two-photon microscopy; (**c**) fluorescence correlation microscopy (FCS); (**d**) fluorescence lifetime imaging microscopy (FLIM).

2.3. Confocal Microscopy

Confocal microscopy limits the observed volume to reduce out-of-focus signals. Minsky introduced the concept of confocal microscopy [19] and issued an original patent for a microscope in 1961. In confocal microscopy, two pinholes that are conjugated in the identical image plane are placed at a focal point in the light path. Light from outside of the focal plane is not focused on the pinhole(s) and only fluorescence very near to the sample's focal point reaches the detector (Figure 2a). Thus, the confocal laser scanning microscopy enables three dimensional reconstruction of specimens.

2.4. Two-Photon Micriscopy

Two-photon microscopy, which was first reported by the Watt W. Webb group in 1990 [20], makes use of the phenomenon that two photons are absorbed by a fluorophore simultaneously. The fluorophore can be excited by light with one-half the energy of each photon or twice the wavelength. The two-photon excitation light is generated by increasing the photon density using a focused high-power femtosecond pulse laser (Figure 2b). Fluorescence emitted from the focus point is detected by a point detector (commonly a PMT) and a fluorescence image is acquired by LSM. Two-photon microscopy enables deep imaging because the microscopy uses near-infrared laser excitation light that exhibits better tissue penetration and collects the localized fluorescence signal.

2.5. Fluorescence Correlation Spectroscopy (FCS)

Fluorescence correlation spectroscopy (FCS) is also based on confocal optics with a continuous wave laser(s) and monitors the mobility of molecules, typically translation diffusion into and out of a small volume (Figure 2c) [12]. FCS analyzes time-dependent fluorescence intensity fluctuations in a tiny observed volume on the order of femtoliter.

When a fluorophore diffuses (or fluorophores diffuse) into the illuminated volume, a fluorescence burst is detected due to steady-state fluorescence emission from the fluorophore(s). If the fluorophore diffuses (or fluorophores diffuse) quickly out of the volume, the burst duration is short, whereas if the fluorophore diffuses (or fluorophores diffuse) more slowly the photon burst duration persists for longer. Correlation analysis of the time series enables the diffusion coefficient of the fluorophore to be determined.

The fluorescence intensity fluctuation depends on the size and number of the molecules passing through the illuminated volume, which provides information on biomolecular interaction in vitro and in vivo.

2.6. Fluorescence Lifetime Imaging Microscopy (FLIM)

Fluorescence lifetime imaging microscopy (FLIM) is an advanced tool that maps the fluorescence lifetime distribution through time-resolved fluorescence detection (Figure 2d) [21,22]. The fluorescence lifetime can respond to changes in pH, temperature, and ion concentrations such as calcium concentration. Its capability to offer both localization of target fluorophores and the fluorophores' local microenvironment exhibits its superiority to fluorescence intensity based steady-state imaging because the lifetime of a fluorophore is mostly independent of its concentration. FLIM can be performed using the time-domain method in which the sample is excited with a pulse laser, or the frequency-domain method in which the sample is excited with intensity-modulated light, commonly sine-wave modulation [12].

3. Single Point Detectors

3.1. Performance Indices

3.1.1. Dark Count

Dark count refers to the tiny flow of electricity in a photodetector operated under a totally dark condition. This dark current should be minimized. Dark count is caused by several phenomena that vary with photodetectors. The dark count of PMTs results from thermionic emission from the photocathode and dynode, and ionization of residual gases (ion feedback) [23]. The dark count of SPADs and SiPMs results from thermionic emission from the depletion layer. The dark count of HPDs is negligible due to its high electron bombardment gain. The dark count of SSPDs, which is low and dependent on the cooling temperature and bias current [24], results from energy dissipation in the nanowire and the blackbody radiation at room temperature through the optical fiber [25].

3.1.2. Instrument Response Function

The transit time and its fluctuation are the major determinants of the time response of a photodetector [23]. The transit time refers to the travel time of the photoelectron. The full-width half-maximum (FWHM) of the instrument response function is a standard for the time response. Many detectors exhibit a non-Gaussian instrument response function. The rise and fall times of a detector are evaluated from the waveform. Transit time spread (TTS) is the fluctuation of the transit time of the single photoelectron pulse.

3.1.3. Afterpulse

Afterpulses are spurious pulses that may appear subsequent to the input signal [23]. An afterpulse is best characterized by the autocorrelation function of the detected photons. For vacuum tube-based photodetectors, positive ions generated by the ionization of residual gases in a detector create afterpulses that appear several hundreds of nanoseconds to several microseconds later than the input signal. This phenomenon is called ion feedback [23]. Among solid-state photodetectors, SPADs and SiPMs experience high afterpulse noise with high count rate measurements. The afterpulse noise is generated by thermally released trapped carriers [26]. HPDs and SSPDs are free of afterpulses.

3.1.4. Sensitivity

The sensitivity of a photodetector is largely determined by the quantum efficiency of the photocathode material. Most photocathodes used in vacuum tube-based photodetectors are made of compound semiconductors. GaAsP exhibits the highest quantum efficiency in the visible region (about 45%). A photocathode (extended red GaAsP) that is more sensitive to a longer wavelength than GaAsP is available for Hamamatsu Photonic products. Silicon and group III-V compound semiconductors, such as InGaAs and GaAs, are also used in solid-state photodetectors. Figure 3 shows the spectral response of these photocathodes.

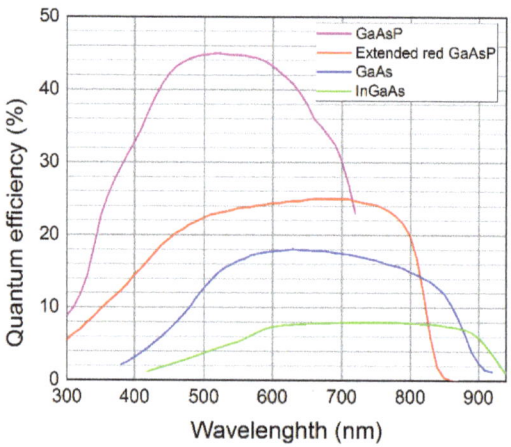

Figure 3. Spectral response of photocathodes.

These photocathodes are usually used in a vacuum and under the ambient temperature ranging from room temperature down to 273 K. As the ambient temperature is lowered, the spectra slightly shift to shorter wavelength due to the bandgap broadening of the semiconductors. Thus, the spectral response of these photocathodes does not change significantly in the temperature range.

3.1.5. Active Area

The diameter or surface of the active area of a photodetector determines the observed area. Confocal optics uses a pinhole in a plane conjugate with the image plane in the sample—the light from the pinhole is easy to focus on a small point detector, such as a SPAD [13]. Point photodetectors with a larger active area are often useful because the light does not need to be focused and they can used to observe a large area without scanning.

3.2. Photomultiplier Tube (PMT)

Photomultiplier tubes (PMTs) are the most widely used point detectors in fluorescence microscopy. A PMT is a vacuum tube that contains a photocathode, focusing electrodes, an electron multiplier (dynodes), and an anode (Figure 4) [23]. Incident photons are absorbed by the photocathode, which ejects primary electrons (~3 eV). The electrons are accelerated by a high voltage to hit a series of dynodes. Then, additional electrons (5–10 electrons) are ejected and exponentially amplified. The electron current is then detected by an external electrical circuit. Typical PMTs have 8–10 dynodes with a cathode-to-anode voltage gap of ~1 kV and current gain of 10^6 to 10^7.

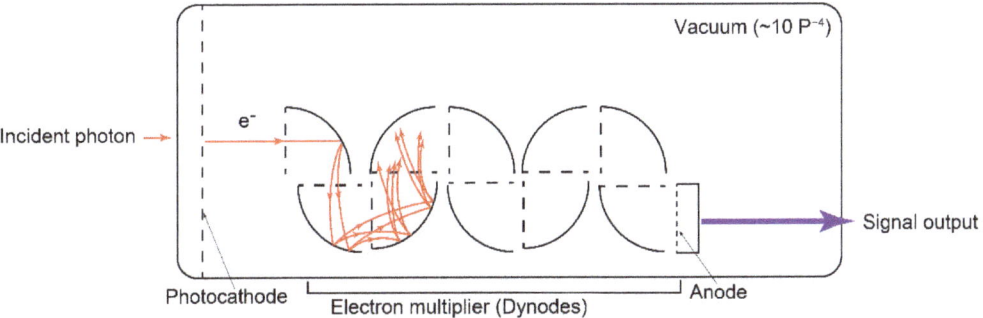

Figure 4. Schematic of a photomultiplier tube (PMT).

The main photocathodes used in PMTs are bi-alkali, with spectral response peaks around 400 nm and range of up to 700 nm. The advent of the GaAsP photocathode with its spectral response peak around 500 nm invigorated the LSM market because its spectral response ranges from visible light (500–700 nm) to near infrared light (900 nm). Laser scanning microscopes with a GaAsP photocathode were released to the market in the late 2000s. Current laser scanning microscopes incorporate PMTs with these multi-alkali photocathodes.

PMTs have been the leading low-light point photodetector for some time and, thus, a large number of power supplies and signal processing circuits for them are available. However, troubles can arise from unwanted noise generated by residual gas molecules (ion feedback), and large variation in responsivity and gain caused by difficulty in controlling metal evaporation to produce photocathodes and secondary electron surfaces. In addition, single photons can be detected with PMTs, but discrimination of single versus multiple photons is difficult.

3.3. Single Photon Avalanche Diode (SPAD)

The single photon avalanche diode (SPAD) is a solid-state photodetector composed of three semiconductor layers, called p-layer, i-layer, and n-layer (Figure 5) [27]. The n-layer has extra electrons, whereas the p-layer has holes. The average gain for an avalanche photodiode (APD) is around 100, which is insufficient for single-photon detection. Therefore, SPADs are usually operated in "Geiger-mode," where an applied bias voltage is greater than the diode's breakdown voltage. Then, when a charge is generated by an incident photon, the charge multiplication (or avalanche) occurs until it saturates corresponding to a current typically specified by the components. These APDs with a single pixel are referred to as SPADs, and those with multiple pixels are referred to as silicon photomultipliers (SiPMs) or multi-pixel photon counters (MPPCs). Although Geiger-mode driven SPADs are suitable for single photon counting, SPADs suffer several drawbacks. The active area cannot be increased because fabrication of a large semiconductor surface increases the number of defects. Geiger-mode driven SPADs have high dark counts due to the Geiger discharge and high afterpulse noise. The dead time is relatively long (about 50 ns), during which photon counting is inoperative because the mode needs to be reset for every single photon detection. To reduce dark counts, SPADs are typically cooled to 210–250 K.

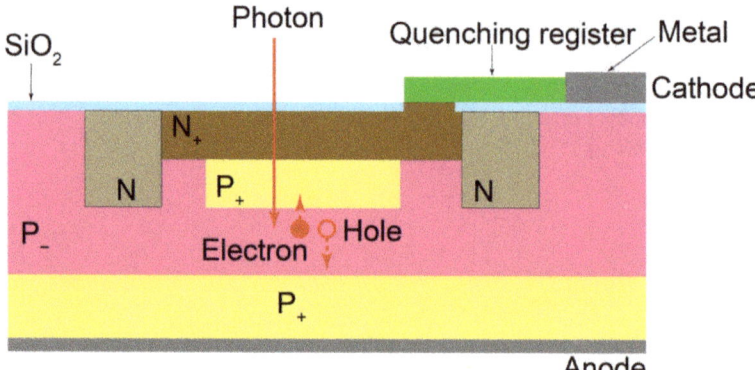

Figure 5. Schematic of a single photon avalanche diode (SPAD).

3.4. Hybrid Photodetector (HPD)

The hybrid photodetector (HPD) is a hybrid of an avalanche diode (AD) and a photocathode, both of which are in a vacuum tube (Figure 6) [23,26,28–31]. When light is incident onto the photocathode, photoelectrons are emitted from the photocathode. The photoelectrons are then accelerated by a high negative voltage to directly bombard the AD where electron-hole pairs are generated and the signal is amplified. The amplification is termed "electron bombardment gain".

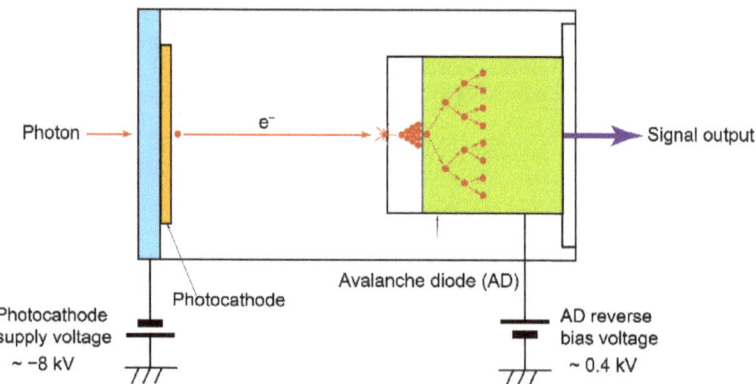

Figure 6. Schematic of a hybrid photodetector (HPD).

In the case of an HPD manufactured by Hamamatsu Photonics, the electron bombardment gain is approximately 1500 with the photocathode supply voltage of −8 kV. The signals of electron-hole pairs are further amplified to 80-fold (avalanche gain) by applying a reverse voltage of about 400 V to the avalanche diode. Then, the total gain will therefore be as much as about 120,000.

3.4.1. Features of HPD

HPDs have significant advantages over PMTs and other low-light photodetectors in the detection of fluorescence, as discussed below. However, a few disadvantages of HPDs also exist. The extremely high cathode supply voltage (−8 kV) is difficult to deal with, which can be problematic when incorporated into systems.

Low Afterpulse

HPDs have a notable feature of lower afterpulse due to their uncomplicated internal structure. Therefore, the major cause of afterpulses, i.e., ion feedback, is highly unlikely to occur in HPDs. Afterpulses evaluated for an HPD and a PMT are shown in Figure 7. This graph shows the probability at which afterpulses may be generated by a single photoelectron input. In contrast to the PMT's multiple afterpulses in a time range from 100 ns to 1 μs, this HPD exhibited only a small number of afterpulses in the time range.

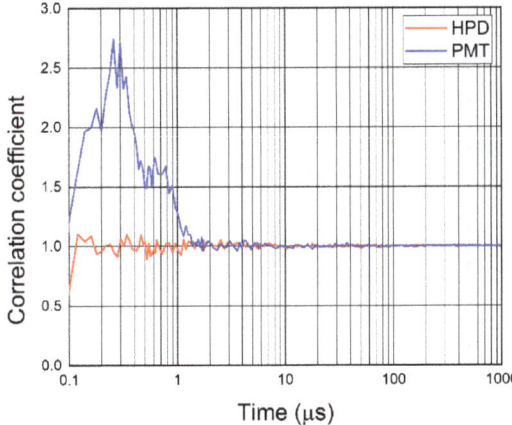

Figure 7. Afterpulse noise of PMT and HPD [32]. The data were obtained through detection of single photons from a continuous wave (CW) light source (wavelength = 470 nm). The cathode supply voltage and the reverse voltage of HPD were −8 kV and 398 V, respectively.

Comparable low afterpulses have been reportedly achieved solely for superconducting nanowire single photon detectors (SSPDs). These SSPDs have micrometer order active areas and must be operated at a liquid helium temperature [25], as discussed in Section 4.2.

High Resolution of Photon Counting

HPDs exhibit better pulse height resolution than PMTs. Gain fluctuation of HPDs is significantly lower owing to much higher electron bombardment gain (about 1500 at a photocathode supply voltage of −8 kV) than the first dynode gain of an ordinary photomultiplier tube (typically as low as 5–10). The first gain mostly determines the signal-to-noise ratio of the electron multiplication, which in turn represents the detector's capability to distinguish between one and multiple photons. As a result, HPDs offer high resolution of photon counting. As shown in Figure 8, signal peaks that correspond to 1, 2, 3, 4, and 5 photoelectrons can be identified in the output pulse height distribution.

High Timing Resolution

Time response characteristics of HPDs are largely determined by the junction capacitance of the internal avalanche diode, provided the diameter of the internal avalanche diode is around or larger than 1 mm. The internal avalanche diode with a diameter of 1 mm and a very low capacitance (4 pF), which is incorporated in the current HPD product, realizes a fast response. Figure 9a shows the time response waveforms of an HPD and a PMT. The FWHM for the HPD is 0.6 ns, which is smaller than that for PMT (1.6 ns).

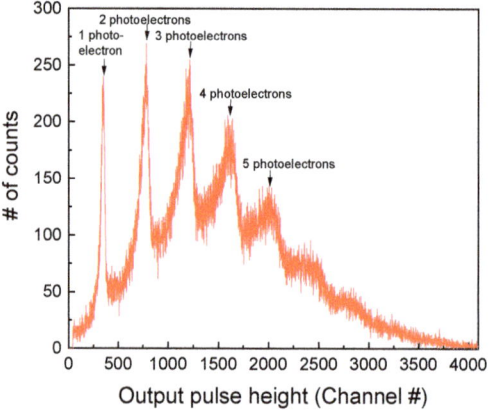

Figure 8. Output height distribution of an HPD [32]. The incident pulsed lights (wavelength = 470 nm) were adjusted to make the photocathode emit three photoelectrons on average. The cathode supply voltage and the reverse voltage were −8 kV and 380 V, respectively.

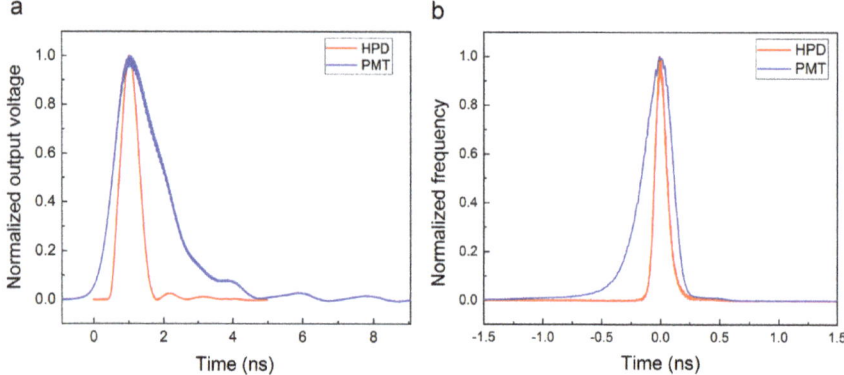

Figure 9. Time response characteristics of an HPD and a PMT [32]: (**a**) time response waveforms of an HPD and a PMT; (**b**) transit time spread (TTS) of an HPD and a PMT. The data were obtained using a pulse laser (wavelength = 405 nm) with a pulse width of 77 ps. The cathode supply voltage and the reverse voltage of HPD were −8 kV and 390 V, respectively.

The TTS determines the instrument response function of HPD. The following three factors mainly affect the TTS for HPD: (i) the transit time within the photocathode; (ii) the variation in the time taken for the photoelectrons to move in the vacuum from the photocathode to the avalanche diode; and (iii) the electron transit time within the avalanche diode. The TTS for an alkali photocathode is about 50 ps [31]. The measured raw TTS for a GaAsP photocathode in an HPD is about 113 ps, which is larger than that for alkali because the GaAsP layer is thicker than the alkali layer. This raw TTS value included the laser pulse width (77 ps) and temporal resolution of the measurement system (30 ps), so the net TTS should be much smaller. Figure 9b shows the time response waveforms of an HPD and a PMT. The raw TTS for the GaAsP photocathode in a PMT is 300 ps, which is larger than that of an HPD. These characteristics are very important for time-resolved fluorescence detection because it contributes to accurate fluorescence lifetime measurement.

Large Active Area

HPDs have a large effective area of more than several millimeters in diameter, enabling high photon collection efficiency. The active area of HPDs is comparable with that of PMTs, which is a marked contrast to SPADs with an effective diameter of only 10 micrometers.

Table 2 lists the characteristics of PMTs, SPADs, and HPDs.

Table 2. Characteristics of point photodetectors [23,27,32].

	Photomultiplier Tube (PMT)	Single Photon Avalanche Diode (SPAD)	Hybrid Photodetector (HPD)
Afterpulse	High	High	Low
Transit time spread (TTS)	~300 ps	~300 ps	~100 ps
Diameter of active area	~5 mm	~several 100 μm	~5 mm
Operating temperature	~270 K	210–250 K	Room temperature

3.4.2. HPD and Fluorescence Microscopy

The features of HPDs mentioned in Section 3.4.1 show the preeminence of HPDs in fluorescence microscopy applications, such as LSM, FLIM, and FCS. In practice, HPDs are incorporated into commercial microscopes and products of Becker & Hickl GmbH and PicoQuant.

FCS

In FCS, the presence of afterpulses deforms the correlation spectra. To avoid this issue, the fluorescence signal is commonly divided into two and detected by two detectors, and the cross-correlation between the two signals is calculated. This procedure is complicated and decreases the signal-to-noise ratio of each acquired datapoint. In contrast, the afterpulse-free feature of the HPD makes FCS measurement simpler, and a single HPD provides better data than PMT [13,26]. As shown in Figure 10, the autocorrelation spectrum acquired by an HPD is of good quality, whereas the spectrum acquired by PMT contains overlapped afterpulse noise below the 2 μs region.

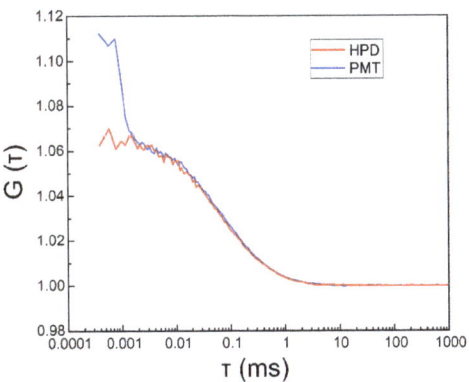

Figure 10. Autocorrelation data obtained by an HPD and a PMT [32]. The data were acquired using 100 nM Alexa Fluor 532 dye solution.

FLIM

FLIM utilizes time-correlated single photon counting (TCSPC) and acquires higher temporal resolution data (<ns) than FCS. Figure 11 shows a comparison of fluorescence life-

time measurement data obtained by an HPD and a PMT. Afterpulses are also problematic for FLIM. The afterpulse raises the baseline of the PMT data. Consequently, the dynamic range of PMT is lower by an order of magnitude than that of HPD and seriously interrupts determination of the decay time constant.

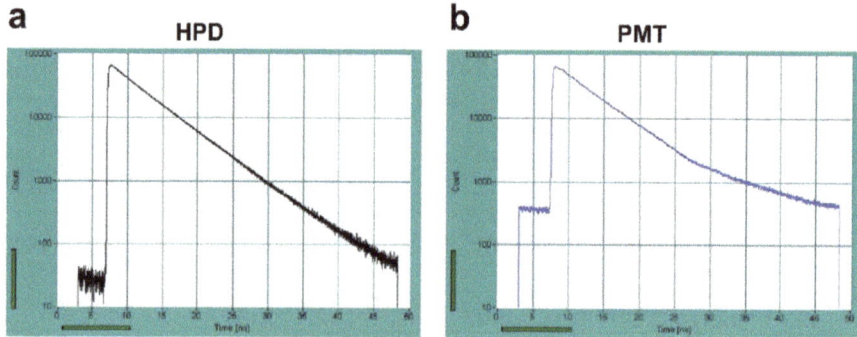

Figure 11. Fluorescence lifetime measurement data obtained by an HPD (**a**) and a PMT (**b**). The data were acquired using fluorescein dye solution. Adapted with permission from [13].

Compared with SPADs, HPDs can obtain brighter FLIM images due to the larger active area of the HPD. Figure 12 shows FLIM images acquired by an HPD and a SPAD. Although the quantum yields of the photodetectors were similar, the image acquired by the HPD collected twice as many photons as acquired by the PMT.

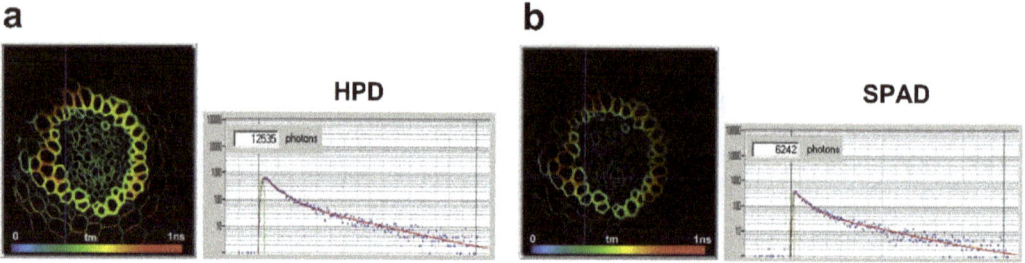

Figure 12. FLIM images acquired by HPD (**a**) and SPAD (**b**). Adapted with permission from [13].

3.4.3. Other Types of HPDs

This subsubsection briefly refers to other types of HPDs that were developed by Hamamatsu Photonics (Table 3) and applications of the HPD to single-molecule fluorescence microscopy. These other types of HPDs are cooled HPD and MPPC (SiPM)-incorporated HPDs.

Table 3. Other types of HPDs [32–35].

Type of HPD	Feature	Possible Application
Photocathode-cooled	Low thermal noise	FCS
MPPC (SiPM)-incorporated	Low voltage operation	Easy to install into systems

Table 3 lists the characteristics of these HPDs.

The cooled HPD, in which the photocathode is cooled by the Peltier element, reduces thermal electronic noise from the photocathode to one-tenth that of the non-cooled HPDs

(Figure 13). This HPD is supposed to be useful for the detection of extremely low light, including single molecule detection.

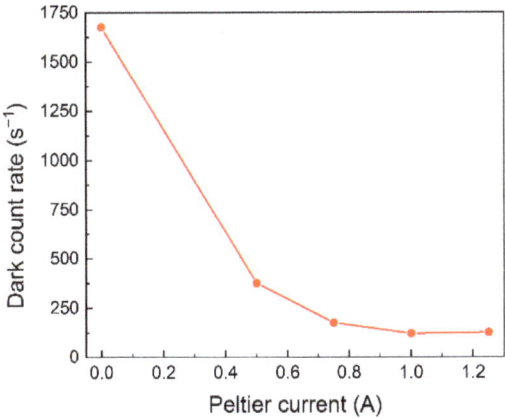

Figure 13. Dependence of dark count rate of the developed cooled HPD on the applied Peltier current [32]. Application of 1.2 A Peltier current cooled the HPD to about 283 K at an ambient temperature of 298 K.

The MPPC-incorporated HPD was developed to solve the difficulties of operating the HPD under high voltages such as 8 kV. The MPPC-incorporated HPD can operate with a lower voltage that operates PMTs due to the MPPC's high gain, and is able to detect single photons. The first prototype of an MPPC (SiPM)-incorporated HPD with a GaAsP photocathode with a diameter of 3 mm and a 25.4-mm (1-inch) bi-alkali photocathode type of MPPC (SiPM)-incorporated HPD were developed by Hamamatsu Photonics, and Barbato et al. evaluated its characteristics [33,34]. As shown in Figure 14, the HPD operates with lower voltages (photocathode voltage ~−3 kV and MPPC bias voltage ~ +70 V) than those of the current HPD product that incorporates an AD (photocathode voltage ~−8 kV and avalanche photodiode bias voltage ~+450 V). The low voltage operation capability facilitates installation of the HPD into various apparatus, including fluorescence microscopes. Recently, Hamamatsu Photonics developed an HPD with a 50.8-mm (2-inch) diameter [35].

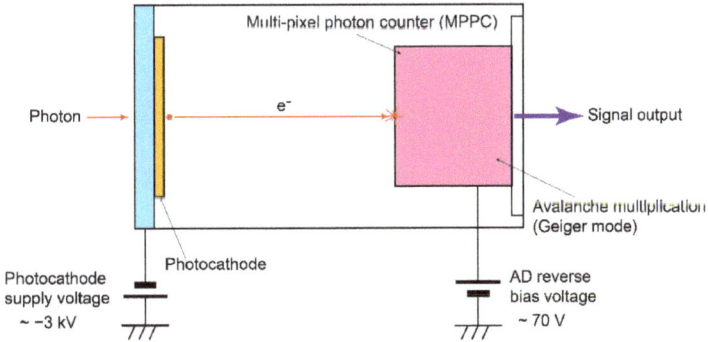

Figure 14. Schematic of multi-pixel photon counters (MPPC)-incorporated HPD.

3.4.4. Application of HPDs to Single-Molecule Fluorescence Microscopy

We applied the cooled-photocathode HPD to single-molecule fluorescence microscopy. We applied the HPD to low-background wide-field single-molecule fluorescence detection with high temporal resolution as a proof-of-principle demonstration. The fluorescence collected by an objective was divided into two, each simultaneously detected by HPD or imaged by EMCCD. The HPD allowed the fluorescence intensity of a mobile single molecule fluorophore to be determined at higher temporal resolution than conventional high-sensitivity CCD cameras. Specifically, the cooled-photocathode HPD detected fluorescence of a single Qdot while performing two-dimensional diffusion with 0.1 ms temporal resolution (Figure 15).

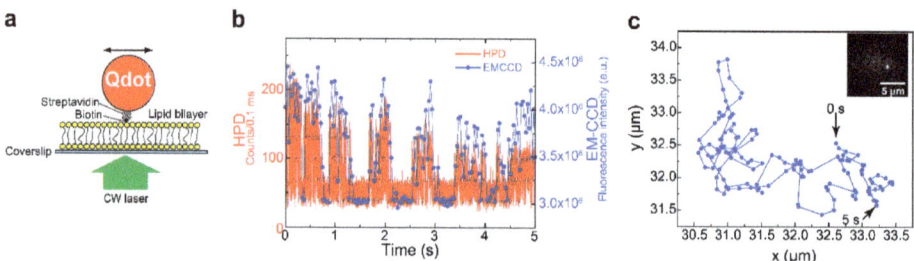

Figure 15. Wide-field sub-millisecond single-molecule fluorescence detection by a cooled HPD [32]: (**a**) Schematic of the observed mobile Qdot. The Qdot (Qdot655 streptavidin conjugate) was attached to biotinylated phosphoethanolamine (PE), 1,2-dioleoyl-sn-glycero-3-phosphoethanolamine-N-(cap biotinyl), via biotin-streptavidin interaction in a 1,2-dioleoyl-sn-glycero-3-phosphocholine (DOPC) lipid bilayer formed on a coverslip. The Qdot was excited by a CW laser (wavelength = 532 nm); (**b**) Time course of a mobile Qdot fluorescence intensity was obtained by an HPD with a time resolution of 0.1 ms (red). The repeated fluorescence-on and –off are caused by blinking. The intensity profile matches the intensity simultaneously obtained by EMCCD (blue) with a time resolution of 31.319 ms; (**c**) Trajectory of the Qdot obtained by EMCCD. The inset is a fluorescence image of the observed mobile Qdot.

The HPD also enabled wide-field single-molecule fluorescence lifetime measurement with nanosecond temporal resolution. We succeeded in obtaining time courses of fluorescence lifetime of a single Qdot whose fluorescence images were simultaneously monitored by an EMCCD (Figure 16).

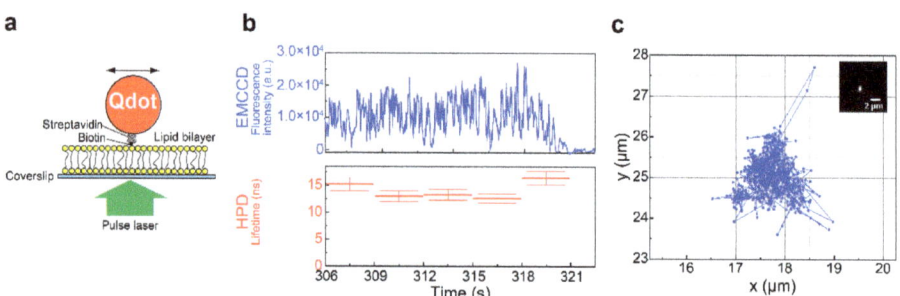

Figure 16. Wide-field single-molecule fluorescence lifetime measurement by a cooled HPD: (**a**) Schematic of the observed mobile Qdot. Sample preparation for the observation was identical to that described in the caption of Figure 15. The Qdot was excited by a pulse laser (wavelength = 520 nm); (**b**) Time courses of the Qdot fluorescence (upper, blue) obtained by an EMCCD and the lifetime (lower, red) simultaneously obtained by an HPD. The lifetime was obtained by fitting each decay curve drawn using data accumulated in three seconds with a single-exponential (red); (**c**) Trajectory of the Qdot obtained by EMCCD. The inset is a fluorescence image of the observed mobile Qdot.

A group at Tohoku University incorporated two HPDs into a specially equipped line confocal optical system in which a slit was used instead of a pinhole to improve the time resolution of single-molecule FRET study. They achieved FRET observation of single protein molecules that flowed unidirectionally in a flow cell at a time resolution of 10 µs and analyzed the high-speed folding process of protein molecules [36].

4. Emerging Point Detectors

This section introduces newly emerging photodetectors that can be used for fluorescence microscopy. These photodetectors are silicon photomultipliers (SiPMs) also known as multi-pixel photon counters (MPPCs) and superconducting nanowire single photon detectors (SSPDs).

4.1. Silicon Photomultiplier (SiPM)

Silicon photomultipliers (SiPMs), which offer single photon detection capability, are an emerging photodetector in a variety of industries and biological fluorescence microscopy. Caccia summarized applications of SiPMs to biophotonics including fluorescence microscopy [37]. SiPM consist of a SPAD array. The sum of pulses from all SPADs is the SiPM output. Note that the number of SiPM output is single irrespective of the number of SPADs in the SiPM. Many photons can be simultaneously detected by the SPADs. Figure 17 shows a schematic of the operating principle of the SiPMs.

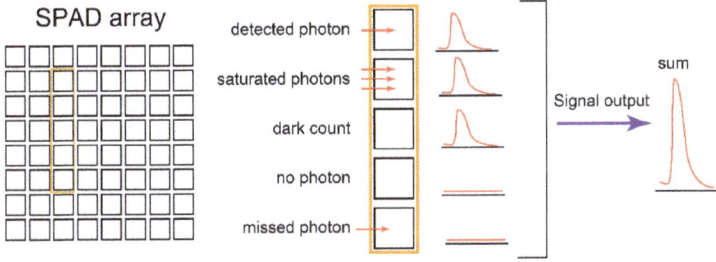

Figure 17. Schematic of the operating principle of silicon photomultipliers (SiPMs): SiPMs consist a SPAD array and every SPAD acts as an element that generates an all-or-nothing current pulse. When one or more photons are absorbed, a current pulse is generated. A current pulse is also produced by the dark count, whereas a current pulse is not generated when photon absorption fails. The output is the sum of the SPADs. Adapted with permission from [38] © The Optical Society.

SiPMs have several advantages over PMTs, including low fabrication cost, low operating voltage, and extremely high damage thresholds (high durability). In addition, silicon diodes have high quantum efficiency in the near-infrared region used for deep tissue imaging. Due to these factors, SiPMs can be better suited for high-speed imaging. Although SiPMs can detect intense light, their dark count rate is larger than that of PMTs, which is the major tradeoff. The advantage of high damage thresholds is reportedly distinct for confocal fluorescence microscopy and two-photon microscopy in clinical sites where surgical marking inks emit intense fluorescence [39]. SiPMs have not yet been widely used for biological imaging. Giacomelli et al. evaluated the performance of commercial SiPMs by comparing the SiPMs with a GaAsP PMT for LSM. They reported that the SiPM sensitivity exceeds the PMT sensitivity for moderate- to -highspeed LSM, whereas the PMT exhibited better sensitivity due to its lower dark counts for low speed LSM [39]. Modi et al. also compared SiPMs products with a GaAsP PMT for two-photon imaging of neural activity [38]. They showed that SiPM exhibited a signal-to-noise ratio that was comparable to or better than PMTs in usual calcium imaging, though dark counts of the SiPMs were higher than that of the PMT. They concluded that the low pulse height variability of the SiPMs surpassed the weak point and resulted in high performance.

4.2. SSPD

In recent decades, the superconducting nanowire single photon detector (SSPD) has become an increasingly popular device, with applications ranging from sensing to quantum communications for single photon detection with high efficiency, precise timing, and low noise [24,40]. The crucial high-speed feature of the SSPD is represented by the TTS (<50 ps) the fluctuation between the true arrival time of a photon and the electrically registered arrival time recorded by the system. Korzh et al. showed that the use of low-latency materials lowered the TTS of the SSPD, and demonstrated that the temporal resolution can be 2.6 ± 0.2 ps for visible wavelengths and 4.3 ± 0.2 ps at 1550 nm using a specialized niobium nitride SSPD [41].

Figure 18 shows a schematic of the SSPD. The SSPD consists of superconducting nanowire with a thickness of a few nanometers that senses photons. Single photon absorption by the SSPD suppresses superconductivity, which in turn generates a voltage spike that can be used to detect the photon.

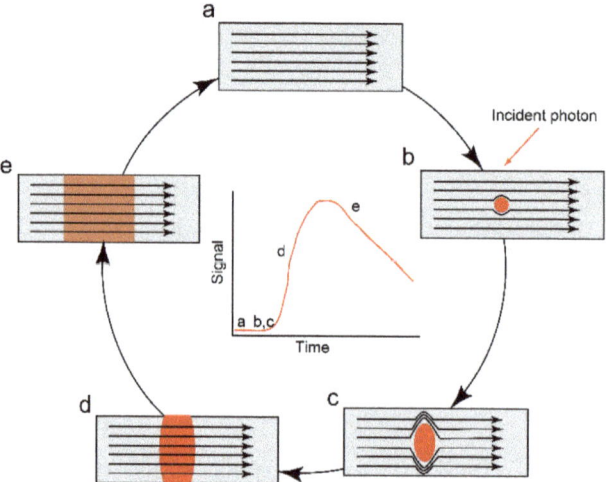

Figure 18. Schematic of the operating principle of the superconducting nanowire single photon detector (SSPD): The SSPD consists of a superconducting nanowire with a thickness of a few nanometers that senses photons. A bias electrical current flows through the nanowire during the operation: (**a**) the entire area is in the superconducting state prior to photon absorption; (**b**) single photon absorption takes place in the superconducting nanowire; (**c**) superconductivity is locally suppressed by energy excitation via the photon absorption; (**d**) the bias current makes the suppressed area resistive and the resistive area expands across the nanowire, which in turn generates a voltage spike; (**e**) as current is diverted, the resistive area relaxes to the superconducting state. Adapted with permission from [42]. Copyright 2020, American Chemical Society.

The SSPD is free of afterpulses because it returns to the superconducting state without generating afterpulses after photon detection. The afterpulsing-free characteristic of the SSPD is a clear advantage for time-resolved fluorescence microscopy. The SSPD has been used for fluorescence microscopy and applied to FCS [43,44]. Although the SSPD has these excellent features, its operation is inconvenient due to its small active area and low operating temperature. The very small active area (~10 µm) makes the optical alignment difficult and the extraordinary low operating temperature ($\leq$4 K) requires the liquid helium cooling system.

Detailed comparisons have been made between SSPDs, SPADs, and other photon-counting technologies in [27,40,44].

The characteristics of SiPMs and SSPDs are summarized in Table 4.

Table 4. Characteristics of emerging photodetectors [24,27].

	SiPM	SSPD
Afterpulse	High	Low
Transit time spread (TTS)	~300 ps	~50 ps
Active area	~3 mm	~10 µm
Operating temperature	250 K–room temperature	≤4 K

5. Summary and Outlook

This paper provides an overview of low-light point photodetectors used in fluorescence microscopy and introduces several point detectors and their operating principles, focusing on HPDs that exhibit high timing resolution and low afterpulse. In addition, we demonstrate application of HPD to wide-field single-molecule fluorescence detection.

Fluorescence imaging with a point photodetector needs scanning to acquire an image and thus temporal resolution of the imaging is often limited. To overcome this limitation, excitation methods other than point excitation, such as multifocal excitation line excitation, have been proposed [45,46]. Regarding photodetectors, imagers with a large number of pixels have been reported by many groups [47–50]. For example, Zickus et al. reported scan-less wide-field FLIM using a camera consisting of a 500 × 1024 SPAD array at a rate of 1 Hz [51]. Michalet et al. firstly reported the evaluation of a multi-pixel (8 × 8) HPD developed by Hamamatsu Photonics [11]. Fukasawa, an author of this paper, and his colleagues, reported another multichannel HPD that was composed of 32 channels (two lines of 16 pixels) on a chip, in which the size of each pixel was 0.8 × 0.8 mm (Figure 19) [52]. It was confirmed that the timing resolution and afterpulse characteristics of the multichannel HPD are identical to the conventional single channel HPD. Wollman et al. reported an 1024-element SSPD array (a 32 × 32 row-column multiplexing architecture) [53]. Fast acquisition methods for FLIM and multi-pixel photodetectors have been reviewed by Liu et al. [22]. These point detector arrays can be widely applied to simultaneous multiparameter observation, including simultaneous multi-wavelength fluorescence observation.

Figure 19. A photograph of the developed multichannel HPD. Adapted with permission from [52]. Copyright 2016, Elsevier.

Low-light photodetectors find more applications in various fields not limited to biological fluorescence microscopy [23]. These applications include flow cytometry and polymerase chain reaction (PCR) in life science, positron emission tomography (PET) for medical diagnosis, elementary particle (neutroino etc.) detection and collision experiments in high energy physics, and semiconductor wafer inspection in industry. In the near future, combination of fluorescence microscopy and other modalities may make low-light photodetectors evolve further and may provide more detailed information on target biological specimens.

Author Contributions: Conceptualization, H.Y.; writing—original draft preparation, H.Y.; writing—review and editing, H.Y., A.F., M.H., and T.I.; investigation, H.Y. and A.F.; formal analysis, H.Y. and A.F.; supervision, H.Y.; visualization, H.Y. and A.F. All authors have read and agreed to the published version of the manuscript.

Funding: This work was supported by the Science Research Promotion Fund of the Promotion and Mutual Aid Corporation for Private Schools of Japan (to H.Y., M.H., and T.I.) and the research grant of Tokai Foundation for Technology (to H.Y. and A.F.).

Data Availability Statement: The data that support the findings of this study are available from the corresponding author upon reasonable request.

Acknowledgments: We would like to thank Yoshihiro Takiguchi for his advice on pulse lasers.

Conflicts of Interest: The author declares that there are no conflicts of interest. The funders had no role in the design of the study; in the collection, analyses, or interpretation of data; in the writing of the manuscript, or in the decision to publish the results.

Abbreviations

AD	avalanche diode
APD	avalanche photodiode
CW	continuous wave
DOPC	1,2-dioleoyl-sn-glycero-3-phosphocholine
EMCCD	electron-multiplying charge-coupled device
FCS	fluorescence correlation spectroscopy
FLIM	fluorescence lifetime imaging microscopy
FRET	fluorescence resonance energy transfer
FWHM	full-width half-maximum
GFP	green fluorescent protein
HPD	hybrid photodetector
LSM	laser scanning microscopy
MPPC	multi-pixel photon counter
PCR	polymerase chain reaction
PE	phosphoethanolamine
PET	positron emission tomography
PMT	photomultiplier tube
Qdot	quantum dot
sCMOS	scientific complementary metal-oxide-semiconductor
SiPM	silicon photomultiplier
SPAD	single photon avalanche diode
SSPD	superconducting nanowire single photon detector
TCSPC	time-correlated single photon counting
TTS	transit time spread

References

1. Lichtman, J.W.; Conchello, J.A. Fluorescence microscopy. *Nat. Methods* **2005**, *2*, 910–919. [CrossRef] [PubMed]
2. Petty, H.R. Fluorescence microscopy: Established and emerging methods, experimental strategies, and applications in immunology. *Microsc. Res. Tech.* **2007**, *70*, 687–709. [CrossRef] [PubMed]

3. Sanderson, M.J.; Smith, I.; Parker, I.; Bootman, M.D. Fluorescence microscopy. *Cold Spring Harb. Protoc.* **2014**, *2014*, pdb top071795. [CrossRef]
4. Moerner, W.E.; Fromm, D.P. Methods of single-molecule fluorescence spectroscopy and microscopy. *Rev. Sci. Instrum.* **2003**, *74*, 3597–3619. [CrossRef]
5. Michalet, X.; Siegmund, O.H.; Vallerga, J.V.; Jelinsky, P.; Millaud, J.E.; Weiss, S. Detectors for single-molecule fluorescence imaging and spectroscopy. *J. Mod. Opt.* **2007**, *54*, 239. [CrossRef] [PubMed]
6. Shashkova, S.; Leake, M.C. Single-molecule fluorescence microscopy review: Shedding new light on old problems. *Biosci. Rep.* **2017**, *37*, BSR20170031. [CrossRef]
7. Yokota, H. Fluorescence microscopy for visualizing single-molecule protein dynamics. *Biochim. Biophys. Acta Gen. Subj.* **2020**, *1864*, 129362. [CrossRef] [PubMed]
8. Betzig, E.; Patterson, G.H.; Sougrat, R.; Lindwasser, O.W.; Olenych, S.; Bonifacino, J.S.; Davidson, M.W.; Lippincott-Schwartz, J.; Hess, H.F. Imaging intracellular fluorescent proteins at nanometer resolution. *Science* **2006**, *313*, 1642–1645. [CrossRef]
9. Huang, B.; Wang, W.; Bates, M.; Zhuang, X. Three-dimensional super-resolution imaging by stochastic optical reconstruction microscopy. *Science* **2008**, *319*, 810–813. [CrossRef]
10. Bertocchi, C.; Goh, W.I.; Zhang, Z.; Kanchanawong, P. Nanoscale Imaging by Superresolution Fluorescence Microscopy and Its Emerging Applications in Biomedical Research. *Crit. Rev. Biomed. Eng.* **2013**, *41*, 281–308. [CrossRef]
11. Michalet, X.; Colyer, R.A.; Scalia, G.; Ingargiola, A.; Lin, R.; Millaud, J.E.; Weiss, S.; Siegmund, O.H.W.; Tremsin, A.S.; Vallerga, J.V.; et al. Development of new photon-counting detectors for single-molecule fluorescence microscopy. *Philos. Trans. R. Soc. B Biol. Sci.* **2013**, *368*, 20120035. [CrossRef] [PubMed]
12. Lakowicz, J.R. *Principles of Fluorescence Spectroscopy*, 3rd ed.; Springer: Berlin/Heidelberg, Germany, 2006.
13. Becker, W.; Su, B.; Holub, O.; Weisshart, K. FLIM and FCS detection in laser-scanning microscopes: Increased efficiency by GaAsP hybrid detectors. *Microsc. Res. Tech.* **2011**, *74*, 804–811. [CrossRef]
14. Michalet, X.; Pinaud, F.F.; Bentolila, L.A.; Tsay, J.M.; Doose, S.; Li, J.J.; Sundaresan, G.; Wu, A.M.; Gambhir, S.S.; Weiss, S. Quantum dots for live cells, in vivo imaging, and diagnostics. *Science* **2005**, *307*, 538–544. [CrossRef]
15. Resch-Genger, U.; Grabolle, M.; Cavaliere-Jaricot, S.; Nitschke, R.; Nann, T. Quantum dots versus organic dyes as fluorescent labels. *Nat. Methods* **2008**, *5*, 763–775. [CrossRef] [PubMed]
16. Kairdolf, B.A.; Smith, A.M.; Stokes, T.H.; Wang, M.D.; Young, A.N.; Nie, S. Semiconductor quantum dots for bioimaging and biodiagnostic applications. *Annu. Rev. Anal. Chem.* **2013**, *6*, 143–162. [CrossRef] [PubMed]
17. Shaner, N.C.; Steinbach, P.A.; Tsien, R.Y. A guide to choosing fluorescent proteins. *Nat. Methods* **2005**, *2*, 905–909. [CrossRef] [PubMed]
18. Shcherbakova, D.M.; Sengupta, P.; Lippincott-Schwartz, J.; Verkhusha, V.V. Photocontrollable fluorescent proteins for superresolution imaging. *Annu. Rev. Biophys.* **2014**, *43*, 303–329. [CrossRef]
19. Minsky, M. Memoir on inventing the confocal scanning microscope. *Scanning* **1988**, *10*, 128–138. [CrossRef]
20. Denk, W.; Strickler, J.H.; Webb, W.W. Two-photon laser scanning fluorescence microscopy. *Science* **1990**, *248*, 73–76. [CrossRef]
21. Becker, W. Fluorescence lifetime imaging–techniques and applications. *J. Microsc.* **2012**, *247*, 119–136. [CrossRef]
22. Liu, X.; Lin, D.; Becker, W.; Niu, J.; Yu, B.; Liu, L.; Qu, J. Fast fluorescence lifetime imaging techniques: A review on challenge and development. *J. Innov. Opt. Heal. Sci.* **2019**, *12*, 1930003. [CrossRef]
23. Hamamatsu Photonics, K.K. *Editorial Committee, Photomultiplier Tubes Basic and Applications*, 4th ed.; Hamamatsu Photonics, K.K., Ed.; Electron Tube Division: Iwata-shi, Shizuoka, Japan, 2017.
24. Natarajan, C.M.; Tanner, M.G.; Hadfield, R.H. Superconducting nanowire single-photon detectors: Physics and applications. *Supercond. Sci. Technol.* **2012**, *25*, 063001. [CrossRef]
25. Yamashita, T.; Miki, S.; Terai, H. Recent Progress and Application of Superconducting Nanowire Single-Photon Detectors. *IEICE Trans. Electron.* **2017**, *100*, 274–282. [CrossRef]
26. Michalet, X.; Cheng, A.; Antelman, J.; Suyama, M.; Arisaka, K.; Weiss, S. Hybrid photodetector for single-molecule spectroscopy and microscopy. *Proc. Soc. Photo. Opt. Instrum. Eng.* **2008**, *6862*. [CrossRef]
27. Eisaman, M.D.; Fan, J.; Migdall, A.; Polyakov, S.V. Invited Review Article: Single-photon sources and detectors. *Rev. Sci. Instrum.* **2011**, *82*, 071101. [CrossRef]
28. Suyama, M.; Kawai, Y.; Kimura, S.; Asakura, N.; Hirano, K.; Hasegawa, Y.; Saito, T.; Morita, T.; Muramatsu, M.; Yamamoto, K. A compact hybrid photodetector (HPD). *IEEE Trans. Nucl. Sci.* **1997**, *44*, 985. [CrossRef]
29. Suyama, M.; Hirano, K.; Kawai, Y.; Nagai, T.; Kibune, A.; Saito, T.; Negi, Y.; Asakura, N.; Muramatsu, S.; Morita, T. A hybrid photodetector (HPD) with a III-V photocathode. *IEEE Trans. Nucl. Sci.* **1998**, *45*, 572. [CrossRef]
30. Fukasawa, A.; Haba, J.; Kageyama, A.; Nakazawa, H.; Suyama, M. High Speed HPD for Photon Counting. In Proceedings of the 2006 IEEE Nuclear Science Symposium and Medical Imaging Conference, San Diego, CA, USA, 29 October–1 November 2006; Volume 1, pp. 43–47.
31. Fukasawa, A.; Haba, J.; Kageyama, A.; Nakazawa, H.; Suyama, M. High Speed HPD for Photon Counting. *IEEE Trans. Nucl. Sci.* **2008**, *55*, 758–762. [CrossRef]
32. Fukasawa, A. Development and Promotion of Hybrid Photodetectors (HPDs) Used in Biological Fluorescence Microscopes. Doctoral Thesis, The Graduate School for the Creation of New Photonics Industries, Hamamatsu-shi, Shizuoka, Japan, 2016.

33. Barbarino, G.; Barbato, F.C.T.; Campajola, L.; Canfora, F.; de Asmundis, R.; De Rosa, G.; Di Capua, F.; Fiorillo, G.; Migliozzi, P.; Mollo, C.M.; et al. A new generation photodetector for astroparticle physics: The VSiPMT. *Astropar. Phys.* **2015**, *67*, 18–25. [CrossRef]
34. Barbato, F.C.T.; Barbarino, G.; Campajola, L.; Di Capua, F.; Mollo, C.M.; Valentini, A.; Vivolo, D. R&D of a pioneering system for a high resolution photodetector: The VSiPMT. *Nucl. Instrum. Methods Phys. Res. A* **2017**, *876*, 48–49.
35. Fukasawa, A.; Hotta, Y.; Ishizu, T.; Negi, Y.; Nakano, G.; Ichikawa, S.; Nagasawa, T.; Egawa, Y.; Kageyama, A.; Adachi, I.; et al. Development of a new 2-inch hybrid photo-detector using MPPC. *Nucl. Instrum. Methods Phys. Res. A* **2018**, *912*, 290–293. [CrossRef]
36. Oikawa, H.; Takahashi, T.; Kamonprasertsuk, S.; Takahashi, S. Microsecond resolved single-molecule FRET time series measurements based on the line confocal optical system combined with hybrid photodetectors. *Phys. Chem. Chem. Phys.* **2018**, *20*, 3277–3285. [CrossRef]
37. Caccia, M.; Nardo, L.; Santoro, R.; Schaffhauser, D. Silicon Photomultipliers and SPAD imagers in biophotonics: Advances and perspectives. *Nucl. Instr. Meth. A* **2019**, *926*, 101–117. [CrossRef]
38. Modi, M.N.; Daie, K.; Turner, G.C.; Podgorski, K. Two-photon imaging with silicon photomultipliers. *Opt. Express* **2019**, *27*, 35830–35841. [CrossRef] [PubMed]
39. Giacomelli, M. Evaluation of silicon photomultipliers for multiphoton and laser scanning microscopy. *J. Biomed. Opt.* **2019**, *24*, 106503. [CrossRef]
40. Hadfield, R.H. Single-photon detectors for optical quantum information applications. *Nat. Photon.* **2009**, *3*, 696–705. [CrossRef]
41. Korzh, B.; Zhao, Q.-Y.; Allmaras, J.P.; Frasca, S.; Autry, T.M.; Bersin, E.A.; Beyer, A.D.; Briggs, R.M.; Bumble, B.; Colangelo, M.; et al. Demonstration of sub-3 ps temporal resolution with a superconducting nanowire single-photon detector. *Nat. Photon.* **2020**, *14*, 250–255. [CrossRef]
42. Allmaras, J.P.; Wollman, E.E.; Beyer, A.D.; Briggs, R.M.; Korzh, B.A.; Bumble, B.; Shaw, M.D. Demonstration of a Thermally Coupled Row-Column SNSPD Imaging Array. *Nano Lett.* **2020**, *20*, 2163–2168. [CrossRef]
43. Yamashita, T.; Liu, D.; Miki, S.; Yamamoto, J.; Haraguchi, T.; Kinjo, M.; Hiraoka, Y.; Wang, Z.; Terai, H. Fluorescence correlation spectroscopy with visible-wavelength superconducting nanowire single-photon detector. *Opt. Express* **2014**, *22*, 28783–28789. [CrossRef] [PubMed]
44. Yamamoto, J.; Oura, M.; Yamashita, T.; Miki, S.; Jin, T.; Haraguchi, T.; Hiraoka, Y.; Terai, H.; Kinjo, M. Rotational diffusion measurements using polarization-dependent fluorescence correlation spectroscopy based on superconducting nanowire single-photon detector. *Opt. Express* **2015**, *23*, 32633–32642. [CrossRef] [PubMed]
45. Yang, W.; Miller, J.E.; Carrillo-Reid, L.; Pnevmatikakis, E.; Paninski, L.; Yuste, R.; Peterka, D.S. Simultaneous Multi-plane Imaging of Neural Circuits. *Neuron* **2016**, *89*, 269–284. [CrossRef]
46. Kazemipour, A.; Novak, O.; Flickinger, D.; Marvin, J.S.; Abdelfattah, A.S.; King, J.; Borden, P.M.; Kim, J.J.; Al-Abdullatif, S.H.; Deal, P.E.; et al. Kilohertz frame-rate two-photon tomography. *Nat. Methods* **2019**, *16*, 778–786. [CrossRef]
47. Colyer, R.; Scalia, G.; Villa, F.; Guerrieri, F.; Tisa, S.; Zappa, F.; Cova, S.; Weiss, S.; Michalet, X. Ultra High-Throughput Single Molecule Spectroscopy with a 1024 Pixel SPAD. *Proc. SPIE Int. Soc. Opt. Eng.* **2011**, *7905*, 790503.
48. Michalet, X.; Ingargiola, A.; Colyer, R.A.; Scalia, G.; Weiss, S.; Maccagnani, P.; Gulinatti, A.; Rech, I.; Ghioni, M. Silicon photon-counting avalanche diodes for single-molecule fluorescence spectroscopy. *IEEE J. Sel. Top Quantum Electron.* **2014**, *20*, 38044201–380442020. [CrossRef] [PubMed]
49. Antolovic, I.M.; Burri, S.; Bruschini, C.; Hoebe, R.A.; Charbon, E. SPAD imagers for super resolution localization microscopy enable analysis of fast fluorophore blinking. *Sci. Rep.* **2017**, *7*, 44108. [CrossRef]
50. Yabuno, M.; Miyajima, S.; Miki, S.; Terai, H. Scalable implementation of a superconducting nanowire single-photon detector array with a superconducting digital signal processor. *Opt. Express* **2020**, *28*, 12047–12057. [CrossRef] [PubMed]
51. Zickus, V.; Wu, M.L.; Morimoto, K.; Kapitany, V.; Fatima, A.; Turpin, A.; Insall, R.; Whitelaw, J.; Machesky, L.; Bruschini, C.; et al. Fluorescence lifetime imaging with a megapixel SPAD camera and neural network lifetime estimation. *Sci. Rep.* **2020**, *10*, 20986. [CrossRef]
52. Fukasawa, A.; Egawa, Y.; Ishizu, T.; Kageyama, A.; Kamiya, A.; Muramatsu, T.; Nakano, G.; Negi, Y. Multichannel HPD for high-speed single photon counting. *Nucl. Instrum. Methods Phys. Res. A* **2016**, *812*, 81–85. [CrossRef]
53. Wollman, E.E.; Verma, V.B.; Lita, A.E.; Farr, W.H.; Shaw, M.D.; Mirin, R.P.; Woo Nam, S. Kilopixel array of superconducting nanowire single-photon detectors. *Opt. Express* **2019**, *27*, 35279–35289. [CrossRef]

MDPI
St. Alban-Anlage 66
4052 Basel
Switzerland
Tel. +41 61 683 77 34
Fax +41 61 302 89 18
www.mdpi.com

Applied Sciences Editorial Office
E-mail: applsci@mdpi.com
www.mdpi.com/journal/applsci

www.ingramcontent.com/pod-product-compliance
Lightning Source LLC
LaVergne TN
LVHW070553100526
838202LV00012B/452